D0845217

Do-It-Yourself Contracting to Build Your Own Home

Also by the author:

Guide to Personal Finance:
    A Lifetime Program of Money Management
Guide to Modern Management (*Forthcoming*)

Do-It-Yourself Contracting to Build Your Own Home

A Managerial Approach

Richard J. Stillman

Chilton Book Company/Radnor, Pennsylvania

For Philip and Grace

Published in Radnor, Pennsylvania, by Chilton Book Company
and simultaneously in Ontario, Canada,
by Thomas Nelson & Sons, Ltd.

Library of Congress Cataloging in Publication Data

Stillman, Richard Joseph
Do-it-yourself contracting to build your own home.

1. House construction–Contracts and specifications.
I. Title.
TH4815.5.S74 1974 643 74-2297
ISBN 0-8019-5937-3

Designed by Cypher Associates, Inc.
Manufactured in the United States of America

Preface

The primary purpose of this book is to inform the person who plans to build a home how to save money and time as well as obtain a better quality residence. Much of the material presented is based upon my own experience living in various parts of the United States, Europe, and the Far East. An extensive bibliography that reflects considerable secondary research is also provided.

For 23 years I was a career Army officer and rented homes in San Francisco; Washington, D.C.; Syracuse; San Antonio; Paris, France; and New Orleans. In each area, I studied the housing market with the intention of one day building my own home. My departure from the military and the assumption of my current job provided this opportunity. As a professor of management, my own home-building experience proved that using sound management principles can save up to 25 percent in costs and accomplish the job in one-third the time. It also gives you a finer quality home because of the personal interest and attention.

This book should also be of considerable value to individuals other than those who wish to subcontract the building of their homes. It can assist people who use an architect and contractor in building a home, as well as all potential home buyers. In each instance, it will help them get a fair deal by providing an understanding of home construction. Furthermore, all home owners who want to add security devices or make other home improvements will find this book an aid.

My own conversations with contractors have shown me that many lack the understanding of management principles needed to build a home efficiently. Most are small-scale entrepreneurs who have no business background. The *Wall Street Journal* ("Out of the Mud") by Jeffrey A. Tannenbaum, June 27, 1972) reports that the building industry "traditionally has been split up among thousands of small firms, usually headed by former carpenters, bricklayers, or other construction workers who made the grade."

Small builders can profit by reading the chapters on planning, organizing, controlling, and managing money. Colleges teaching management, business, and/or public administration courses could find this book useful as a case study. A number of schools are using such books to illustrate management principles. It is an effective technique in my classes.

If you decide to subcontract your home, keep in mind that it is a challenge. But the rewards can be king size. There is the psychological satisfaction of actually participating in the building of your dream home. There is also the economic reward of savings in time, money, and materials. For those who accept the challenge, please let me know the results.

Cristine L. Persons and Benton Arnovitz are my editors and deserve a special thank you.

My thanks to the Stillman family who had to put up with a difficult author. Darlene, my wife, provided valuable administrative assistance. Tom and Ellen are teenagers and it was no small challenge to keep the peace. Richard, our older son, is an author and California State professor of public administration. He offered words of encouragement and made an important home repair contribution.

To Helen and Roy Fisher, my appreciation for many years of wise counsel in the financial management aspects of housing. Elizabeth G. Kramer has been my indispensable secretary for the last three years. She was responsible for the final typing and editing. Dora Lee Hall has been a most loyal and conscientious helper.

This book has been published thanks to many nice people who have been very cooperative. This includes advice from subcontractors, bookstore managers, bank and savings and loan association officials, business executives, consumer organizations, colleagues, students, and friends. In particular, I wish to mention Weston Strauch, Bob Ruby, Frank McGuire, B. F. Doss, Honorable Joe W. Sanders, Dr. Martin Klein, Dalton L. Woolverton, Merlin Toups, and Walter C. Balser.

Now comes the obvious: For the undiscovered errors, necessary improvements, and such, the entire responsibility is mine. Please let me know of desired changes so the needed modifications may be made in a subsequent revision.

Richard J. Stillman
New Orleans, La.

Contents

Do-It-Yourself Contracting to Build Your Own Home

CHAPTER 1
A Managerial Approach to Building Your Home

Introduction

This chapter presents an overview of the entire book. It is devoted to an understanding of management principles and how you can apply them to subcontract your own home.[1] Throughout the following chapters, we will use these concepts to point out how it is feasible to manage the building of your house in order to save both money and construction time.

By using this approach, I saved 25 percent on the cost and completed my home in one-third the normal time for a comparable residence. Furthermore, the quality of materials and workmanship were superior because of the personal attention I devoted to it.

Definition and Model

What is management? It means achieving an objective by utilizing men, money, and material; performing the functions of planning, organizing, and controlling; working within a framework of line, staff, and service responsibilities; and becoming involved in the decision-making process.

Let's look at a management model (Figure 1-1) that will help to examine this definition. In addition, an example is provided that will enable you to relate the management model to your homebuilding project. The model presents a graphic overview of the five major management components: objective, resources, functions, areas, and decision-making process.

Objective

Management must establish a primary goal for its organization. Assume that a homebuilder decides his objective is to build a residence in a designated area,

[1] These same management principles have universal application and can be utilized at work, in a social organization, and in managing your money.

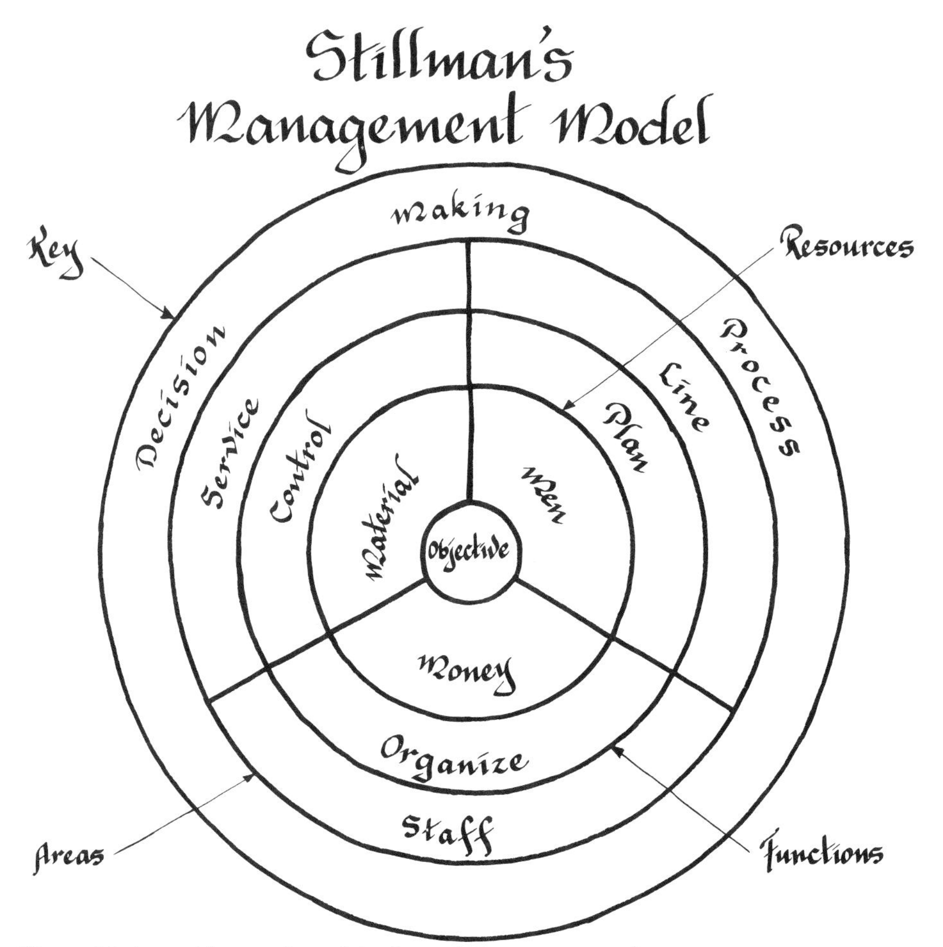

Figure 1-1 A graphic overview of the five major components of management.

utilizing subcontractors and completing the project within two months at a 25 percent savings in total cost.[2]

The objective should provide answers to the questions of what, when, where, how, and why. Figure 1-2 presents these answers as they pertain to a homebuilder.

Establishing a primary objective also means you'll need to develop sub-objectives or sub-goals. The more complex the homebuilding project, the more involved these sub-goals become. They may also include short, medium, and long-term objectives. As the model makes clear (Figure 1-1), the objec-

[2] The objective must be realistic and feasible. It should be adopted only after you've given appropriate thought to available resources (men, money, material) and the other components of the management model.

tive is at the heart of management. All management efforts should be focused on the accomplishment of the objective.

Resources

Once your management objective is established, you must determine what resources are required to most efficiently accomplish your objective. In order for any organization to operate effectively, it requires adequate manpower, financial resources, and equipment.

Manpower would require one full-time individual (you or your wife) who has overall responsibility. In addition, this individual might draw upon forty-nine different subcontractors from time to time, as well as inputs from a lawyer and several qualified technicians in the homebuilding industry. Money power would be needed to pay the subcontractors and all other expenses. Material power involves lumber, roofing, concrete, sheetrock, and many other supplies. In essence, the three inputs of men, money, and material result in one output: a completed home.

Functions

The third circle in the management model refers to the three basic management functions: planning, organizing, and controlling. In order for a manager to effectively accomplish his objective, he must be capable of performing these three functions:

1. Plan what you are going to do.
2. Do what you plan.
3. Check what you have done.

	Home Case
WHAT	Build a residence
WHEN	Within two months
WHERE	In a designated area
HOW	Subcontracting
WHY	Save 25%

Figure 1-2 The management objective should provide answers to the questions of what, when, where, how, and why.

Later chapters will emphasize the importance of using a budget to serve as a *planning* document and as a means of *control* to determine how the actual expenditures compare with the planned estimate. The actual building of the house encompasses the *organizational* or doing aspect. These three management functions apply in every type of homebuilding, from the smallest cottage to the largest mansion.

Areas of Modern Management

Now that we have established our objective, determined our resources, and recognized our functions, let us look at the areas of responsibility that may be utilized to accomplish our objective. The homebuilder must decide what *line*, *staff*, and *service* areas are most appropriate. *Line* activities would include such production area subcontractors as carpenters, plumbers, electricians, and roofers. These people are the doers and comprise the operational aspect of management.

In addition to the line requirement, there is a need for *staff* responsibilities. These are the activities that help the line people or operators accomplish their various assignments. The homebuilder would perform the vast majority of the staff work himself: keep the financial records, serve as his own personal manager, and provide his own research. However, he would hire a lawyer for legal work and rely upon his wife or helper for clerical and administrative tasks.

The third area of responsibility may be categorized as *service* activities. These include custodial, maintenance, and storage duties. This service function, performed by the owner or a member of his family, includes sweeping the premises, running errands for the subcontractors, carrying refuse to the dump, and performing a multitude of other menial but vital tasks.

Decision-Making Process

The final ring portrayed in Figure 1-1 is the decision-making process. The decision-making process may draw upon all other components of the management model in order to arrive at a sound solution. In making management decisions, these questions should be kept in mind:

1. What is your objective?
2. Do you have the necessary facts to make a sound decision?
3. What are your alternatives?
4. Have you chosen the most profitable alternative?

Guidelines

During the planning and building phases, you may also wish to consider the following guidelines which are highlighted throughout the book.

1. Time is a valuable commodity; employ it to your advantage. A management approach to homebuilding minimizes the time required to complete the project.
2. Be flexible and imaginative. In our dynamic society, a home should be attuned to the times, present and future. This includes consideration of the fact that one day you may wish to sell your house.
3. Make your homebuilding project a family affair by encouraging active participation of all members.
4. Go first class. The higher-priced materials and quality workmanship cost less in the long run and give greater satisfaction.
5. Maximize your down payment; contract for the shortest term mortgage you can afford; and pay off the balance as rapidly as possible—if you have found the locality where you expect to settle for at least three years.
6. Minimize your costs by eliminating the middleman wherever possible.
7. Deal only with reputable people who have a good record of achievement and stand behind their work.
8. Pay all bills by check. But keep an adequate amount of money in your checking account to allow you to secure the best prices from "hungry" subcontractors or suppliers who prefer "cash on the barrel head" rather than waiting 30 days for their money. The remainder should be either drawing interest or borrowed only at the time required.
9. Your home should be a major growth investment. The lot, house, and maintenance should be viewed in this perspective.
10. During the planning and construction, capitalize on your family's experience.
11. Timing is important. This applies to availability of materials, men, and money as well as weather conditions.
12. Don't let the magnitude of this homebuilding project frighten you. Look at the various tasks and break each down to its smallest component. From such a perspective, each job becomes relatively easy.
13. Make every effort to get along with people, including suppliers, subcontractors, other builders, inspectors, bank officials, and neighbors.
14. It pays to be in good physical and mental condition since you may be working 10 to 12 hours each day and will find the task physically and mentally challenging.

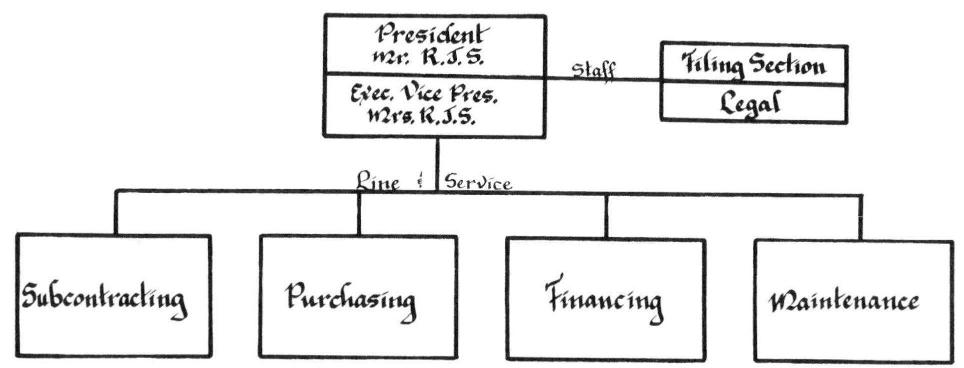

Figure 1-3 A hypothetical organization to manage construction of a home.

Organization

One approach to organizing your homebuilding project effectively is to consider yourself the business manager with overall responsibility (see Figure 1-3).

If you undertake this project as a single person, there isn't any question who is boss; but marriage may present a different story. Accordingly, you may wish to have co-presidents rather than the arrangement in this chart.

Let's assume you designate yourself as president and your wife agrees to be your executive vice president—with authority to act in your behalf if you should be absent. This organizational system (Figure 1-3) should be developed during the initial planning stage. Be specific in delineating responsibilities but make your organization flexible so changes can be made if necessary. You should have a workable filing system and a person to handle all correspondence; my wife performed this function admirably. She also was responsible for the purchasing of material. We used a young lawyer friend for our legal matters. In addition to overall responsibility, I made the financial arrangements and was boss of each subcontractor—plumber, electrician, carpenter, roofer, bricklayer, and so on. Our son took charge of the maintenance chore, but we all pitched in when necessary.

Why a Managerial Approach to Homebuilding?

A home is usually the largest single investment a person makes in his lifetime. Therefore, it is very important to examine every aspect thoroughly. This need for a careful study is supported by comments obtained in a large metropolitan area where homes varied in cost from $20,000 to $100,000. In a majority of cases, the owners stated that major pitfalls could have been avoided if greater care had been taken prior to purchase. Here are the remarks of two dissatisfied homeowners that typify the magnitude of problems people face when they fail to utilize sound management concepts.

> We bought our house the way we would buy a pair of shoes. Looked at the house one afternoon and within an hour signed an agreement. It's been a financial nightmare. Our monthly payments, closing costs, utilities, and maintenance expenditures have exceeded by 40 percent what the salesman estimated. My wife had to get a job and we are still going further in debt. We can't sell at anywhere near the price we paid. How stupid of us not to have taken our time and studied up on what we were getting into.

The following remarks were based on an interview with an owner of a $75,000 home in an exclusive neighborhood.

> A friend suggested I contact his cousin to build my home. The guy seemed pleasant and had built a few houses in the city. But it's been a damn nightmare. From the start, he alienated everybody. He made the neighbors mad by trampling their shrubbery with his trucks, using their water without permission, and throwing cement and other debris on their lawn. He fought every suggested change I made and then charged us twice the going price. He had no idea of what communication or management was all about. For example, we asked for a fine quality redwood fence. Instead, I received a tan color of cheaper grade. His excuse: "I told the subcontractor what to do but he screwed up." I ended up painting it red. As you can imagine, it looks terrible.
>
> The house was to be finished in six months and it took ten. Many things didn't work when I moved in. He promised to fix everything, but I ended up paying to have someone else do it. My wife and I spent an excessive amount of time supervising work that our builder was supposed to do. But he hardly ever showed up. To make matters worse, the plumber and others wouldn't take instructions from us so we wasted more time trying to contact him. It was a mess. The house has nearly caused my wife a nervous breakdown. She is still on tranquilizers.

Summary

In conclusion, management principles can be applied to subcontracting your own home and can help you save both money and time. The management model presents a graphic overview, consisting of five components that can be helpful in arriving at sound decisions. Familiarize yourself with administrative guidelines to assist in the planning and building phases. Likewise, an organization chart can be helpful in delineating responsibilities.

Now that we have an overview of a managerial approach to building your home, let us examine a specific area of the model—the vital planning phase.

CHAPTER 2
Planning Your Dream House

The first management function is adequate planning. In order to be a successful homebuilder, you must spend sufficient time during the planning phase. Adequate planning means determining if this is the correct time to build: read relevant available literature;[1] speak with building experts; and observe quality, residential construction firsthand. Also give serious consideration to location, design, expenditures, time available for the project, and selection of subcontractors and desired materials. If this planning goes well, all other requirements should fall into place. My wife and I spent eleven months doing our preparatory work.

Timing—Should I Build Now?

During this period of inflation and high interest rates, I am frequently asked, "Is it wise to build at this time?" My answer is, "Yes." If you intend to live in a community for at least three years, you should do well by making such an investment. Prices of housing materials and labor are going up each day and there is no indication that inflationary pressures will not continue in the future, although the rate may be somewhat slower than the 1973-1974 rise. Nevertheless, it will hit hardest at the pocketbooks of fixed income people; and even if the rates are lowered, they will be reduced in a lesser proportion than building costs will rise.

Another aspect you should consider is availability of subcontractors and suppliers. If there is a major building boom in your area, it could delay your

[1] The federal government has good documents available. Pointers can be obtained from the FHA publication, *Minimum Property Standards for One and Two Living Units*, FHA No. 300, November 1966, with revisions; and the Department of Agriculture Bulletin No. 168, "House Construction (How to Reduce Costs)," August 1969. Also read articles in the bibliography that may be pertinent to your project.

completion by months. The large contractors would normally tie up the best subs and some materials might not be in local stock. Recently I talked with a small builder who was most unhappy:

> I have waited 12 weeks for my plumber to return and install the bath fixtures—excuses, excuses, excuses. Last year he would have come begging, but now he has so much work he claims to be working 14 hours a day including Sunday. This delay has already cost me $4,000. I am paying 8 percent on the $150,000 I borrowed for these four duplexes. Vandalism and theft have been terrible. As you can see, holes have been punched in my celotex, my pipes have been broken, paneled doors kicked in, sliding glass partitions jammed, and a small fire destroyed part of one house. Stupidly, I didn't have insurance to cover the fire loss. To make matters worse, some of my good lumber was carted away and 500 of my bricks disappeared. I will be lucky to break even on this job.

Timing is also important because of seasonal weather conditions. In some areas, floods, typhoons, heavy snows, drenching rains, excess heat, hard freezes, or hurricanes can play havoc with house construction. You must also determine when it would be most convenient for you and/or your wife to supervise the project. In most areas, it is helpful to get the actual building underway in the late spring. The longer days are advantageous to a homebuilder. Your subs work their people on an hourly wage and are delighted to put in long hours to finish their work and get paid. This helps you in your objective to complete the house on schedule.

PERT and Timetable

An important planning tool I used is a modified PERT (Performance Evaluation and Review Technique) concept. PERT provides a technique for "planning the step-by-step development and completion of a project."[2] It is a valuable tool to "help identify potential bottlenecks and determine ways of choosing between the resultant delays or alternative approaches that will reduce delays."[3]

We used PERT for both the planning and building phases (see Figure 2-1). Shortly after our arrival in New Orleans, I developed a planning schedule

[2] Haynes and Massie, *Management: Analysis, Concepts, and Cases*, 2d ed. (Englewood Cliffs, N.J.: Prentice-Hall, Inc. 1969), p. 263; also note, pp. 570-74. (Courtesy of Prentice-Hall, Inc.)

[3] Ibid., p. 263.

that seemed feasible: a nine month schematic diagram that was an optimistic approach and an alternative twelve month schedule, allowing for unplanned incidents. The detailed schedule, much like a budget, provides both a plan and a control technique.

A timetable listing events in chronological order can also be of valuable assistance in checking on your progress. It requires that you list all major steps in planning your home. A model timetable is shown in Figure 2-2.

Location

Location is of prime importance. The following features should be considered: distance to work; transportation facilities; taxes; people currently living in the residential area; availability of police and fire departments; proximity of good schools if you have or plan to have children; suitability of land to provide relative safety, e.g., high ground will provide relative safety during a flood; area where property values are rising.

It may be wise to limit your search to within a 25 mile radius of your employment. With the ever-increasing traffic problems, a premium should be placed on proximity to work. Make certain to determine the travel time between home and office.

Today, more people are searching for a house within walking, jogging, or biking distance. This makes good sense. It saves time and money, improves your health, and contributes to the ecology movement.

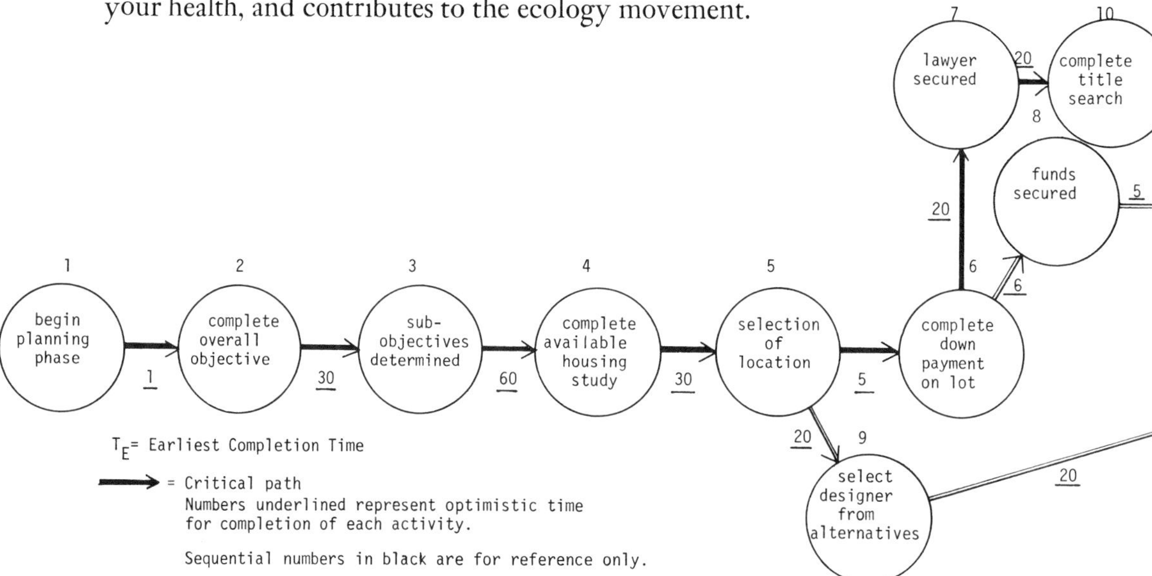

Figure 2-1 PERT chart used during the planning phase of the Stillman house.

Real estate taxes, present and planned, must be considered in determining where to live. A recent article pointed out that in one large community the city dweller would fare better than those in suburbia. Mr. James Maloney, president of the Real Estate Board of New Orleans, said:

> The trend is not to add real estate taxes, not to burden the city homeowner. The people who buy houses know what the taxes are going to be and we can live with the tax base the way it is. But the outlying areas have expanded so rapidly that a new tax base may be required. The gravy train is over concerning taxes in suburban spots and everyone is not happy living in suburbia. More and more they will be attracted to the city.[4]

Take the time to look at alternative lots. Speak to neighbors in the immediate area and consider the privacy factor, if it's of importance to you. Give consideration to the homes adjacent to your lot. Will neighbors be looking in your backyard even with a six-foot fence?

The area you select should have a strong neighborhood association that checks to be sure homes meet minimum specifications. Pride in neighborhood appearance should be encouraged.

[4] Frank L. Schneider, "Return to the City Beginning Here? Some Feel It May Be," New Orleans Times-Picayune, February 18, 1973, p. 17. (Courtesy of Times-Picayune Publishing Corporation)

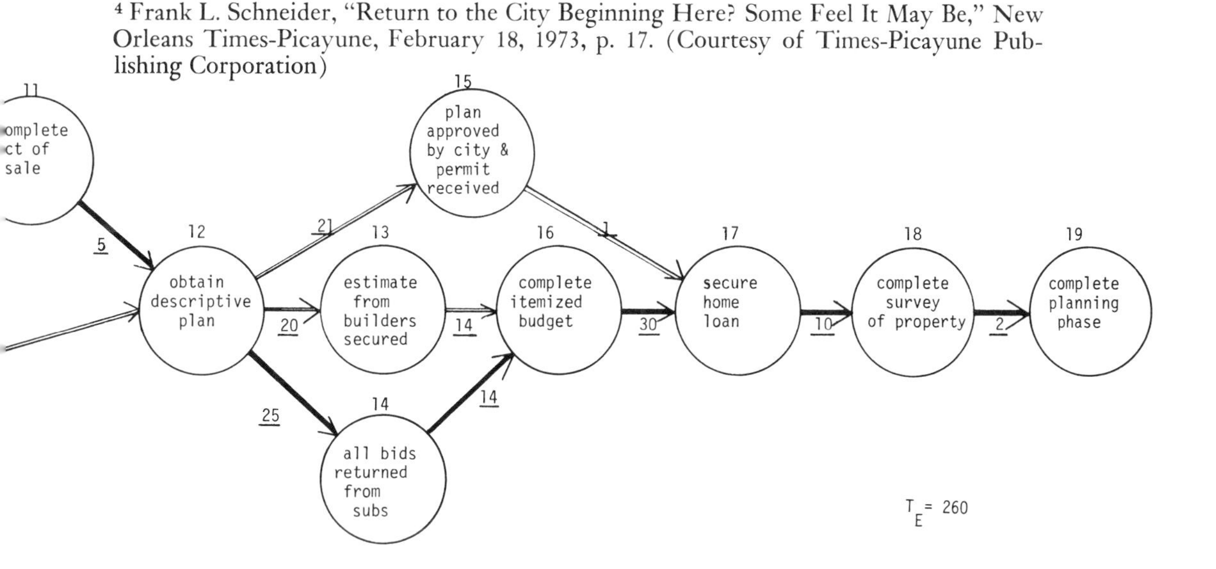

1. Begin planning phase. Discuss techniques to meet needs—PERT, management model, planning steps.
2. Overall objective—a home within one year to meet family requirements and within financial resources.
3. Subobjectives:
   a. Location
      1. Walking distance to work
      2. Professional residential area
      3. Near good schools
      4. High ground to provide relative safety in the event of hurricane
      5. Area where property values are rising
   b. Cost
   c. Type
4. Complete study as to available housing:
   a. Read articles
   b. Obtain ideas from authorities
   c. Visit sites
5. Select from alternatives the desired location (i.e., 10 to 15 lots)
6. Complete down payment on lot
7. Secure lawyer
8. Secure funds for lot
9. Select designer from alternatives
10. Complete title search
11. Complete act of sale
12. Obtain description of materials and necessary plans
13. Secure estimate for comparison from builders
14. All bids returned from subs
15. Plan approved by city and permit received
16. Complete itemized budget
17. Secure home loan
18. Complete survey of property
19. Complete planning phase—review

Figure 2-2 A model timetable for a managerial approach to homebuilding: Planning phase.

Building Restrictions

Building restrictions established by the city and the subdivision should be studied very carefully. Be sure to obtain the most current copy and check with the local association president to see if there are plans to modify any of the present restrictions.

The rules regarding home construction may include height of the house; advance approval of the plans; garbage receptacle standards; minimum front, side, and rear yards; instructions for services such as gas, telephone, and elec-

	A	*B*	*C*	*D*	*E*	*F*	*G*	*H*	*I*
Proximity to work									
Convenience to school									
Convenience to church									
Play area nearby									
Quality of police and fire protection									
Adequacy of zoning restrictions									
Convenience to shopping facilities									
Location in relation to factories									
Size of lot in relation to price									
Drainage									
Total									

SCALE

Outstanding	4 points
Good	3 points
Fair	2 points
Poor	1 point

Figure 2-3 Ratings for various lots under consideration for purchase.

tric power; grading of site; restrictions on hedges and shrubbery; fences; and swimming pool standards. Figure 2-4 lists the building restrictions established for one area.

Purchase Arrangements

The next step is to make arrangements for purchase of the desired lot. Hopefully you can buy direct from the owner and he will negotiate. If not, you should work through an able real estate agent (member, local real estate board). This will permit you to arrive at a fair price; fair in comparison to similar lots recently sold. (The sale prices of all lots appear weekly in some local papers.) Prior to hiring a lawyer, secure his itemized charges. (See Figure 2-5 for sample charges.)

Title Search and Insurance

You should have a title search or title insurance. However, if there have been only one or two previous owners, it may not be necessary.

The purpose of a title search is to determine if the lot is clear of prior claims. Your lawyer will usually check the records back for 50 to 75 years for this purpose. You can expect to be charged one-percent of the property price for this service. Thus, on a $10,000 lot it will cost you $100. Make certain you get what you pay for. That is, be sure the attorney spells out in a letter the length of time the title search covered.

BUILDING RESTRICTIONS FOR
LAKE OAKS SUBDIVISION
A – Single Family Dwelling Sites

These restrictions apply to Squares No. 3, 4, 5, 6, 7, 8, 9, 10, 11, 5087, 5088, 5089 and 5090 in "Lake Oaks Subdivision", which is in Zone Four of the Lakefront Development of The Board of Levee Commissioners of the Orleans Levee District, in the City of New Orleans, Louisiana. Said subdivision is bounded by Lake Pontchartrain on the north, property of American Radiator & Standard Sanitary Corp. and St. Roch Ave. on the east, various properties on the south and Elysian Fields Ave. on the west.

SECTION I
DEFINITIONS
MAP

Where reference is made to the Map of "Lake Oaks", it refers to the Map entitled, "LAKE OAKS ON LAKE PONTCHARTRAIN, NEW ORLEANS, LOUISIANA, BOARD OF LEVEE COMMISSIONERS OF THE ORLEANS LEVEE DISTRICT," dated December 1, 1960, File No. L. D. 3106, signed by A. L. Willoz, C. E. Registration No. 73.

ORLEANS LEVEE BOARD

Where the name "Orleans Levee Board" appears it shall mean "The Board of Levee Commissioners of the Orleans Levee District."

MAIN BUILDING

The building that will be inhabited, containing rooms such as Living Room, Kitchen, Dining Room, Parlor, Bedroom, Library, etc.

ACCESSORY BUILDING

A subordinate building, attached to or detached from the main building, the use of which is incidental to that of the main building and not used as a place of habitation, such as a Living Room, Kitchen, Dining Room, Parlor, Bedroom, Library, Bathroom, etc.

HEIGHT OF BUILDING

The height of a building is the perpendicular distance measured in a straight line from the top of the highest point of the roof beams in the case of flat roofs, and from the average height of the gable in case of a roof having a pitch of more than twenty degrees with a horizontal plane, downward to the established grade in the center of the front of the building.

SECTION II
APPROVAL OF BUILDING PLANS

Prior to beginning the construction of a residence, garage, swimming pool, fence, or other structure, the owner shall submit reproduced copies of detailed plans and specifications of the proposed building or structure to the Orleans Levee Board. With plans the owner shall have to furnish an affidavit, certifying that the value of the house shall be equal or above the minimum in Section XVI. No work shall be done on the building until such written approval is received and building permit obtained from the City of New Orleans.

The approval of all structures by the Orleans Levee Board will be based on the requirements of these restrictions and on appearance.

Owners are cautioned that all structures erected on any parcel of ground in Lake Oaks must comply to Zoning Ordinances of the City of New Orleans. There may be cases where the City Ordinances are more restrictive than these title restrictions, in which case the former will govern.

SECTION III
GENERAL

All lots in Squares No. 3, 4, 5, 6, 7, 8, 9, 10, 11, 5087, 5088, 5089, and 5090, of Lake Oaks Subdivision shall be devoted to single family dwellings. On these lots all the usual uses normally allowed to private homes such as by professional men who operate Dentists' and Doctors' Offices, Nurseries, Clinics, etc., therein will not be permitted.

SECTION IV
FRONTAGE

No residence shall be built on less than one lot as shown on Map of "Lake Oaks". When any purchaser wishes to buy more than one site in order to erect a larger permitted residential building, this may be done provided that said lots or fractional lots shall aggregate a width of at least as wide as adjacent lots and shall be treated as one lot. The restrictions applying to a single lot should apply to such larger site. No resubdivision of lots shall be done which would leave remaining on the square a lot of an area or width below the average standard for said square, as indicated on the Map of "Lake Oaks". No lot shall be shifted as to frontage. No lot shall be renumbered or lose its identity even when subdivided. No lot shall be resubdivided for private sale or otherwise unless first approved by the Orleans Levee Board.

HEIGHT

Residences shall not exceed thirty-five (35) feet in height.

SECTION V
MINIMUM FRONT, SIDE AND REAR YARDS

(a) No part of any building of a residence shall be built closer than twenty (20) feet minimum distance from the front property line of the lot, nor closer to either side property line of the lot than six (6) feet minimum distance, however, the combined width of both side yards shall be a minimum of twenty-five per cent (25%) of the width of the lot, but need not exceed twenty-one (21) feet. Corner Lot 21, Sq. 4, Lots 1 and 20, Sq. 5 and Lot 19, Sq. 5087 shall have twenty (20) feet set backs on front and side streets. All other corner lots shall have twenty (20) feet set back on front street and twelve and five tenths feet (12.5) on side street. No detached garage shall be constructed closer than sixty (60) feet from the front property line and when outside the required rear yard area shall not project into the side yards.

An attached or detached carport may be constructed in the side yard area, provided its columns or wall are not closer than six (6) feet from the side property line, however, the roof may project two (2) feet into six (6) feet area. An attached or detached carport shall be considered a building for storing automobiles and having no more than one side enclosed. The other three sides must be completely open. Storage areas, utility rooms, tool rooms, etc., in such carports will not be permitted within the side yard area. Cooling towers and condensers must be erected in the rear of the main building and not project into side yard areas. No cooling towers or condensers shall be erected on roof of any building. When an accessory building is outside the required rear yard area it shall not project into the side yards. Clothes lines shall be in the rear of main building and shall not project beyond the side building lines.

(b) Bay or bow oriel, dormer and other projecting windows, stairways, landings or other structural parts shall not project beyond the front and side building lines.

(c) Cornices, roof overhangs, gutters, spouting, chimneys, brackets, pilasters, grill work, trellises, and other similar projections and any projections for purely ornamental purposes may project beyond the front and side building lines, however, not exceeding two (2) feet.

(d) Unenclosed, uncovered or covered porches, balconies and steps, shall not project beyond the front or side building lines.

(e) The rear yard measured from the farthest back projection of the principal building to the rear property line shall be not less

Figure 2-4 Sample building restrictions established for one area.

In order to avoid or reduce this cost, you may wish to check the records yourself or obtain the services of a title searcher who specializes in this work. His fee should be no more than half the lawyer's charge. In fact, some lawyers use the services of these title searchers and then charge the client the full

than twenty per cent (20%) of the depth of the lot, except that in deep lots said yard need not exceed at any point, a maximum of twenty-five (25) feet and on shallow lots no rear yard shall be less, at any one point than a minimum of fifteen (15) feet.

(f) The front of a corner lot shall be its narrower dimension.

SECTION VI
VEHICLES

No trucks, trailers, automobiles or other commercial vehicles bearing advertisements are to be stored or parked on residential property or on streets, except when making deliveries. Passenger vehicles, owned by a resident, shall be stored on the resident's ground and not on the street.

SECTION VII
GARBAGE RECEPTACLES

Each residence of Lake Oaks Subdivision shall be required to install a garbage receptacle between the front sidewalk and curbing, by the owner of said residence. This receptacle shall be of sufficient capacity to take care of the garbage cans used by said residence and shall be of an approved design. The receptacle shall be installed underground and shall have a neat cover flush with the sidewalk surface. Details of garbage receptacle shall be made part of the plans and specifications of the residence to be submitted to The Board of Levee Commissioners of the Orleans Levee District for approval. No garbage cans are to be exposed on the street or sidewalk in front of a residence.

SECTION VIII
SERVICES

All services, such as gas, telephone, electric power, sewers, drains and water pipes shall be placed underground from the property line to the building.

Relative to electric service, the owner shall lay, or have laid, a cable underground from his meter to a New Orleans Public Service Inc's., transformer vault in the rear of the lot.

Relative to telephone service, the owner shall provide, at his own risk and expense, an open trench not less than fifteen (15) inches in depth from his house to a telephone terminal box in rear of his property. Location of this trench is to be designated by the telephone company. The Southern Bell Telephone and Telegraph Company will then lay the necessary cable in this trench and the owner, after the cable is laid, will then backfill this trench at his own expense.

SECTION IX
TRANSFORMER VAULTS AND EASEMENTS

To serve the sites in the Lake Oaks Subdivision with an electric underground system, transformer vaults will be erected in the rear of lots adjoining an easement owned by the Orleans Levee Board within which the primary cables serving the transformers will be located. These vaults will be constructed of brick or concrete and will be located as indicated on Map of Lake Oaks Subdivision, File No. L. D. 3106.

Public easements to permit walking through from street to street are provided in Squares 7, 8, 9, 10 and 11, and also between Squares 5 and 6. These easements are indicated on Map of Lake Oaks Subdivision File No. L. D. 3106.

On the lots listed below there are indicated on the plan of Lake Oaks four (4) foot easements, which are reserved for underground electric conduits and cables to be installed and maintained by the New Orleans Public Service Inc., as part of the Electrical Distribution Systems serving the residences of Lake Oaks Subdivision. The conduits will be encased in concrete and will be at least three (3) feet below ground surface which will permit landscape plantings and erection of permissible structures under Section V.

Sq. 3	Lot 2
Sq. 4	Lot 5
Sq. 4	Lot 24

Although the probability is remote, repairs to the conduits may become necessary, and in such event, the New Orleans Public Service Inc., has the right to enter these easements to make the necessary repairs.

When such repairs have been completed, the New Orleans Public Service Inc., will be responsible to restore the surfaces of the easements to their original condition upon completion of said work.

SECTION X
SWIMMING POOLS

Swiming pools, if and when erected are to be of approved, substantial and neat construction, and will only be permitted provided they are entirely surrounded by a fence not less than thirty-six (36) inches in height and shall conform to all fence requirements recited in Section XII. The vertical inside faces of a pool shall be built no closer than six (6) feet to either side property line nor closer than thirty (30) feet to the front property line, not closer than six (6) feet to the rear property line. The finished top-side or surface deck shall not be constructed higher in elevation than one foot above the established site grade of the residence. Equipment such as diving boards, etc., shall not be higher than five (5) feet above site grade of the residence. The pool shall be so designed as to prevent splashings from the pool from draining into adjacent properties.

SECTION XI
PARKS

Lake Oaks Park is a public park, owned by the Orleans Levee Board, and is for use of the General Public of the City of New Orleans. This park will be maintained by the Orleans Levee Board.

Interior Parks are reserved for the common use of the property owners of Lake Oaks and nothing shall be placed thereon, or no use shall be made thereof, to the detriment, inconvenience or annoyance of the resident, or owner of any part or portion of ground adjacent thereto. These parks are owned and are to be maintained by the Orleans Levee Board.

SECTION XII
FENCES

Fences will be permitted as noted below:

Front yard fences, if and when erected, shall not exceed eighteen (18) inches in height and shall be of neat and substantial construction.

Side fences, when erected between the front building line and front property line, shall not exceed eighteen (18) inches in height.

Side and rear yard fences, if and when erected between front building line and rear property line, shall not exceed five (5) feet in height and must be of neat and substantial construction.

On corner lots, fences shall not be erected closer than the required setback from any street or park property line.

Plans showing location and details of fences must be submitted for approval to the Orleans Levee Board before they are erected.

SECTION XIII
GRADE OF SITE

The established grade of lots is not to be raised by any individual owner so as to adversly affect an adjacent property owner or owners in the same square.

SECTION XIV
PLANTING

Hedges and shrubbery may be grown along property lines, but shall be restricted to a height of two (2) feet along the front yard property line, and shall be restricted to a height of seven (7) feet on side and rear property lines.

Trees planted in the required side yards of one lot may not project into the required side yard of adjacent owners, except upon agreement between the affected owners that said projection is not objectionable.

All trees, shrubbery, flowers, lawns or other vegetation on private residential lots shall be kept in good order by the owners and/or their tenants.

SECTION XV
GRASS CUTTING ON VACANT SITES

For the purpose of keeping the Subdivision in an orderly condition, the Orleans Levee Board reserves the right and assumes the responsibility to cut the grass on vacant building sites for the period of time commencing after the improvements are completed to December 31, 1962. In cutting the grass on vacant property the Orleans Levee Board is to be held harmless for any damage by the owners of such vacant property. Upon termination of this period it will be the responsibility of the owners of each lot to maintain the grass in a presentable condition.

SECTION XVI
MINIMUM COST OF RESIDENCES

The total construction cost of any residence shall be at least equal to or in excess of one and one-half (1½) times the market value of the lot, on which constructed, at the time construction begins.

Figure 2-4 Sample building restrictions established for one area.

price. See Figure 2-6 for an example of a letter spelling out the details of a title search that you should request from your lawyer.

Title insurance provides you protection in the event it is later found that you don't really own the property. The charge is normally one percent of the

EXPENSES:	*BUYER*	*SELLER*
PAYABLE AS COURT COSTS:		
Mortgage certificate		5.00
Paving certificate		2.00
Conveyance certificate		3.50
Tax certificate		4.00
Registration of act in conveyance office	2.25	
Registration certificate	.75	
Cancelling mortgage Registration in archives, New Orleans		2.25
PAYABLE AS LEGAL FEES:		
Passing act of sale	60.00	
Preparing, ordering, and obtaining certificates		25.00
Conducting title examination	25.00	
Executing release of mortgage		25.00
Less: deposit on costs	−30.00	
PAYABLE AS OTHER COSTS:		
PRO RATA TAX ADJUSTMENT:		
City taxes (to buyer)	118.62	
State taxes (to seller)		10.30
PURCHASE PRICE:		
Close out balance on loan with first homestead (to homestead)	11,385.29	
Seller's portion (to seller)	4,814.71	
TOTAL DISBURSEMENTS	$16,375.62	$77.05

Figure 2-5 Itemized legal charges regarding our lot.

land and home. For example, if this amounts to $40,000, your title insurance may cost $400. Title companies specialize in this type of business. Obviously, if your attorney has done a thorough title search, there should be no necessity for title insurance. Unfortunately, this may not happen in cases involving frequent changes of ownership over a long period of time.

Title insurance protects the lender only and not the purchaser. The purchaser, however, can protect himself by paying an additional sum of money. Be sure to check this out prior to making a decision. If you let your lawyer take care of title insurance, he may charge an extra amount or receive a kickback from the company selected.

If the records indicate you have clear title to your lot, you then sign the "agreement to purchase" (see Figure 2-7).

Plans and Materials

Keep your costs for plans at a minimum. If you can find a young man—not necessarily licensed but competent—who has worked for an architect and has just started his own business, he will be delighted to prepare your plans for a reasonable fee.

We developed our own sketches of the house, as well as detailed specifications. Our draftsman used the sketches to make up ten sets of drawings (five plans per set: foundation, floor plan, elevation, plot plan, typical interior and exterior wall sections). It is important to work closely with this individual in order to make certain your desires are conveyed clearly to him and that your proposals are feasible. You can keep your draftsman happy and keep your costs down by not making changes after his drawings have been finalized. (See Appendix 1 for a complete set of drawings.)

I am amazed at the number of people who buy a home and never ask to see the drawings. They provide valuable information that could be an important factor in arriving at your decision to buy or not to buy. (This also applies to the "description of materials.")

The draftsman or architect should also provide you with ten copies of "description of materials." It is important to be specific about materials for

Dear Dr. Stillman:

The writer has examined the title to the above described property, and based upon the records and documents available for inspection, the title is approved subject to the certificates obtained and copies of survey. Certificates have been received clear of any discrepancies in prior transfers dating back through two preceding transferors.

Mortgage certificates indicate that there is presently an outstanding mortgage and vendor's lien in the amount of $11,385.29 in favor of 1st Homestead and Savings Association, such mortgage being evidenced by a note executed by Anthony L. Guardina in favor of the above mentioned mortgagee on December 20, 1965.

This open vendor's lien in favor of 1st Homestead and Savings Association, recorded in M.O.B. 2093, folio 60, of Orleans Parish, will be cancelled subject to payment of the outstanding balance of $11,385.29 to the vendor, 1st Homestead and Savings Association.

The last survey of this property was conducted more than two years ago, and it is suggested that an up-to-date survey be secured prior to construction.

Regular city and state tax certificates and special paving, sewerage, and water certificates have been secured and indicate that payment of all taxes has been accomplished. City and State property taxes, it is understood, will be prorated among the vendor and vendee as indicated in the Act of Sale.

Yours very truly,

Figure 2-6 Title examination by our lawyer.

AGREEMENT TO PURCHASE OR SELL

For exclusive use of "Realtors"

FORM ATP-OS NO 3
Revised and approved January, 1971
Standard Form
Real Estate Board of New Orleans, Inc.

Member Real Estate Board of New Orleans, Inc.

________ ____________________ Agent ____________________, La., ____________________ 19____

1. ____________ offer and agree to sell / purchase ____
2. ____
3. ____
4. On grounds measuring about ____ or as per title.
5. Property sold and purchased subject to all title and zoning restrictions on record, or by laws or ordinances for the sum of ____________________
6. ____ ($ ____________) Dollars,
7. on the terms of ____________________________ cash ____________________________
8. ____
9. ____
10. This sale is conditioned upon the ability of purchaser to borrow upon this property as security the sum of $ ____________________ by a mortgage loan or
11. loans at a rate of interest not to exceed ____________ % per annum, interest and principal payable on or before ____________________ years
12. in equal monthly installments.
13. Should purchaser, seller or agent be unable to obtain the loan stipulated above within ____________ days from acceptance hereof, this contract shall then be
14. null and void and the agent is hereby authorized to return the purchaser's deposit in full. Commitment by lender to make loan, subject to approval of title, shall constitute ob-
15. taining of loan.
16. Property sold subject to the following lease or leases:
17. Tenant ____________ Rental ____________ Expiration ____________ Options ____________
18. ____________ ____________ ____________ ____________
19. Possession and occupancy by purchaser at passing of act of sale or on ____________________________
20. All recorded sewerage, utilities and street surfacing charges bearing against the property as of this date to be paid by ____________________
21. Real Estate Taxes and rentals (if any) to be prorated to date of act of sale.
22. All costs and fees for necessary certificates and vendor's closing fee to be paid by seller. Cost of survey by ____________________
23. Act of sale at expense of purchaser to be passed before ____________________________ Notary, at any time after
24. ____________________ 19____, and prior to ____________________ 19____. In the event bona fide curative work in connection with title
25. is required, the parties herewith agree to and do extend the time for passing of act of sale by thirty days.
26. Upon acceptance of this offer, vendor and purchaser shall be bound by all its terms and conditions and purchaser becomes obligated to deposit with seller's agent immediately
27. ____________ % of the purchase price amounting to $ ____________, and failure to do so shall not void this agreement but shall be considered as a breach
28. thereof and seller shall have the right, at his option, to demand liquidated damages equal to the amount of the deposit or specific performance, and purchaser shall, in either
29. event, be liable for agent's commission, attorney's fees and costs.
30. This deposit is to be non-interest bearing and shall be placed in any bank in Metropolitan New Orleans, without responsibility on the part of the agent in case of failure or
31. suspension of such bank.
32. The seller shall deliver to purchaser a merchantable title, and his inability to deliver such title within the time stipulated herein shall render this contract null and void,
33. reserving unto purchaser the right to demand the return of the deposit from the holder thereof, and reserving unto agent the right to recover commission.
34. In the event the seller fails to comply with this agreement for any other reason, within the time specified, the purchaser shall have the right either to demand the return of
35. his deposit in full plus an equal amount to be paid as penalty by the seller; or the purchaser may demand specific performance, at his option.
36. In the event the purchaser fails to comply with this agreement within the time specified, the seller shall have the right to declare the deposit, ipso-facto, forfeited, without
37. formality beyond tender of title to purchaser; or the seller may demand specific performance.
38. In the event the deposit is forfeited, the commission shall be paid out of this deposit, reserving to the seller the right to proceed against the purchaser for the recovery of
39. the amount of the commission.
40. If this offer is accepted, seller agrees to pay the agent's commission of ____________________ which commission is
41. earned by agent when this agreement is signed by both parties and when the mortgage loan, if any, has been secured.
42. Either party hereto who fails, for any reason whatsoever, to comply with the terms of this offer, if accepted, is obligated and agrees to pay the agent's commission and all
43. reasonable attorney's fees and costs incurred by the other party, and/or agent in enforcing their respective rights.
44. ____
45. This offer remains binding and irrevocable through ____________________________
46. Submitted to ____________________ (Signed) ____________________
    Listing Agent or Owner
47. ____________________
    Address
48. By ____________________
    Selling Agent
    I/We accept the above in all its terms and conditions.

____________________, La. (Signed) ____________________

____________________ 19____ ____________________

____
Address

Figure 2-7 Sample "Agreement to Purchase or Sell." (Courtesy of Real Estate Board of New Orleans, Inc.)

every item in your home—foundation, exterior walls, floor framing, roofing, windows, interior doors and trim, cabinets, heating, plumbing, and so on. For example, for roofing specify "asbestos shingles, Johns-Manville®, american colonial." Don't simply ask for "asbestos shingles," because there is a great deal of quality and price difference in the variety of roofing. It is your responsibility to provide the draftsman with the specific details on material. If undecided, indicate a well known item and add "or equivalent." Being as precise as possible will permit more accurate estimates by builders and subcontractors. (See Appendix 2 for a sample description of materials.)

Our experience indicates that it pays to buy the finest rather than mediocre materials. In the long run, the cost differential is recovered in durability, reduced heating and cooling bills, and aesthetic satisfaction. The preparation of a description of materials may sound like a big undertaking, but this is not so. Your designer will probably have many sets from previous work, and good builders also have copies. The owner of a home you admire may provide you with his plans. Once you have several descriptions of materials in hand, you need only make appropriate modifications to suit your needs.

Builders and Estimates

After the plans and description of materials are completed, you can let out three sets to respected builders in the community. They will give you their prices to construct this exact house. You will then have a comparison of their costs versus your costs. Keep in mind that if you want any additions after signing with a builder, they will cost you extra.

I checked with fourteen individuals who had contracted to have their homes built and found that the original estimate had been increased by 8 to 26 percent because of additional charges the builders had made. One of my neighbors had a patio overhang extended four feet—it cost him an extra $800.

Budget

Bear in mind other costs such as landscaping, fence, and drapes. Therefore, when you estimate your budget, every cost you can think of should be itemized in the estimate, including your moving expense. Itemize every cost in your budget; it is a significant planning tool. I devised the format shown in Figure 2-8. You may be able to obtain a form for this purpose from a lending agency. I recommend that you devise a budget to best suit your needs.

Bids

Once you have listed all the items necessary to build your home, it is time to contact subcontractors and supply houses for prices. Leads on the most reliable firms can be obtained from (1) successful builders, (2) observation of

The Stillman Budget

Item	*First Estimate*	*Revised Estimate*	*Actual*	*Contracted With:* (Name, address, phone, principal to contract)
Preliminary Costs [Includes such items as lot, taxes, insurance, lawyer's fee, building permit, and survey.]				
Foundation:				
Piling (#9)	950.00	912.00	912.00	Smith Pile Drivers, Inc.
.				121 Richton Ave.
.				711-3000
.				Mr. John Smith
Materials				
Lumber	3,579.79	3,641,50	3,766.14	Doe Lumber Company
.				5164 Collins Avenue
.				401-6111
.				Mr. J. J. Doe
Total *	xxx	xxx	xxx	

* In arriving at your total, you may wish to pay yourself a fair return and include this figure in the overall price. My economic opportunity cost meant that if I hadn't spent twelve hours per day on this job for two months, I would have done other productive work. However, if I had contracted with a builder there would also be time expended. A query of four persons indicated they averaged three hours per day over an eight-month construction period. Buying a new or used home also involves use of time—loan negotiations, etc.

Figure 2-8 The Stillman itemized budget.

quality homes in progress and talks with the subcontractors concerned, and (3) friends who did their own subcontracting. All three sources should prove valuable.

If feasible, get bids from three sources of suppliers and subcontractors before you make your final decision. We did not always take the lowest bid or accept the first price. In order to secure bids on lumber, electrical requirements, plumbing, and so on, it is necessary to provide these firms with your plans and description of materials. This is the primary reason for having ten sets made. Be sure to have copies returned. (The city and local jurisdictional board may also require copies for their files.)

Once you have selected the best deal, it is only a matter of tabulating the prices to determine the overall cost of building your home. This total can be compared with what the three builders would have charged. You may find a 25 percent differential with even the lowest bid. If the spread is close or the project appears insurmountable, you may wish to go the builder route—or buy a home. But don't be discouraged at this point!

Summary

Adequate planning is essential to build your home economically and efficiently. Let's assume you have arrived at a decision to subcontract your home. You have done a good job of planning with respect to selecting the location; obtaining appropriate plans; preparing a budget, timetable, and PERT chart. The budgeted costs have been arrived at based upon appropriate bids. The next chapter will look at ways to secure the needed money—the ingredient that makes the dream a reality.

CHAPTER 3

Money— The Essential Ingredient

Now that you have done a good job of planning, and have all your facts in hand, you are in a position to visit lending agencies. Officials at banks, mortgage firms, insurance companies, and savings and loan associations should be impressed with your knowledge and the thoroughness of your work. Check your budget once more to be sure that all estimated expenses are listed and that your computations are correct. Go over your description of materials to see that each item is specified.

Your objective now is to obtain the money under the best possible terms. The loan you seek to build your home is called a mortgage. Title or ownership of the home will rest with the lending agency until you have paid your debt to that institution (see Figure 3-1 for a sample mortgage). According to Webster, a mortgage is "a conveyance of property upon condition that becomes void upon payment or performance according to stipulated terms." You should become thoroughly familiar with the terms of your mortgage *before you sign it.* A careful reading of Figure 3-1 will help acquaint you with the terminology.

Interest Costs

One of the most important considerations will be the interest you must pay and the number of years you will be making payments. Even ¼ percent can be a significant amount. For example, a $40,000 mortgage points up the cost spread based upon different interest rates of varying duration:

	PERCENT							
Years	7	7¼	7½	8	8½	9	9½	10
5	$ 7,523.00	$ 7,806.80	$ 8,091.20	$ 8,663.50	$ 9,240.20	$ 9,820.40	$10,404.80	$10,993.40
10	15,732.80	16,353.20	16,977.20	18,238.40	19,514.00	20,805.20	22,112.00	23,433.20
15	24,717.20	25,727.00	26,745.80	28,808.60	30,902.00	33,027.80	35,184.20	37,373.00
20	34,428.80	35,878.40	37,337.60	40,299.20	43,311.20	46,376.00	49,486.40	52,642.40
25	44,816.00	46,739.00	48,680.00	52,619.00	56,630.00	60,704,00	64,844.00	69,047.00
30	55,806.80	58,236.80	60,688.40	65,663.60	70,725.20	75,866.00	81,086.00	86,370.80

Sale of Property
BY
Greater New Orleans
Homestead Association
TO

UNITED STATES OF AMERICA

STATE OF LOUISIANA

CITY OF NEW ORLEANS

BE IT KNOWN, That on this

day of , in the year nineteen

hundred and

Before me,
a Notary Public in and for the City of New Orleans, in the Parish of Orleans, and State of Louisiana, aforesaid, duly commissioned and qualified, and in the presence of the witnesses hereinafter named and undersigned,

PERSONALLY CAME AND APPEARED:

herein acting for and representing the

GREATER NEW ORLEANS HOMESTEAD ASSOCIATION

a duly incorporated institution of this State, domiciled in this City, created by act before John Janvier, Notary, dated June 19, 1909, recorded in the Mortgage Office of this Parish in Book 950, Folio 297 et seq., amended by an act passed before Jacob D. Dresner, Notary Public, on September 17, 1924, recorded in Mortgage Office Book 1303, Folio 456, also amended by an act passed before Jacob D. Dresner, Notary Public, on June 17, 1926, recorded in Mortgage Office Book 1338, Folio 141, and amended by an act passed before Jacob D. Dresner, Notary Public, on October 13, 1937, recorded in Mortgage Office Book 1536, Folio 354, herein and hereunto authorized by a Resolution of the Board of Directors of said Association, adopted at a meeting of said Board, a copy of which resolution is hereto annexed for reference and made part hereof, which said appearer declared that for the consideration and on the terms and conditions hereinafter expressed, he does, by these presents, in said capacity, grant, bargain, sell, convey, transfer, assign, and set over unto

here present, accepting and purchasing for heirs and assigns, and acknowledging delivery and possession thereof, the following described property, together with all the buildings and improvements thereon, and all rights, ways, privileges, servitudes, appurtenances, and prescriptions, liberative and acquisitive, thereunto belonging or in anywise appertaining, to-wit:

Figure 3-1 A sample mortgage form for the state of Louisiana. (Courtesy of M. R. Toups, Executive Vice President, Greater New Orleans Homestead)

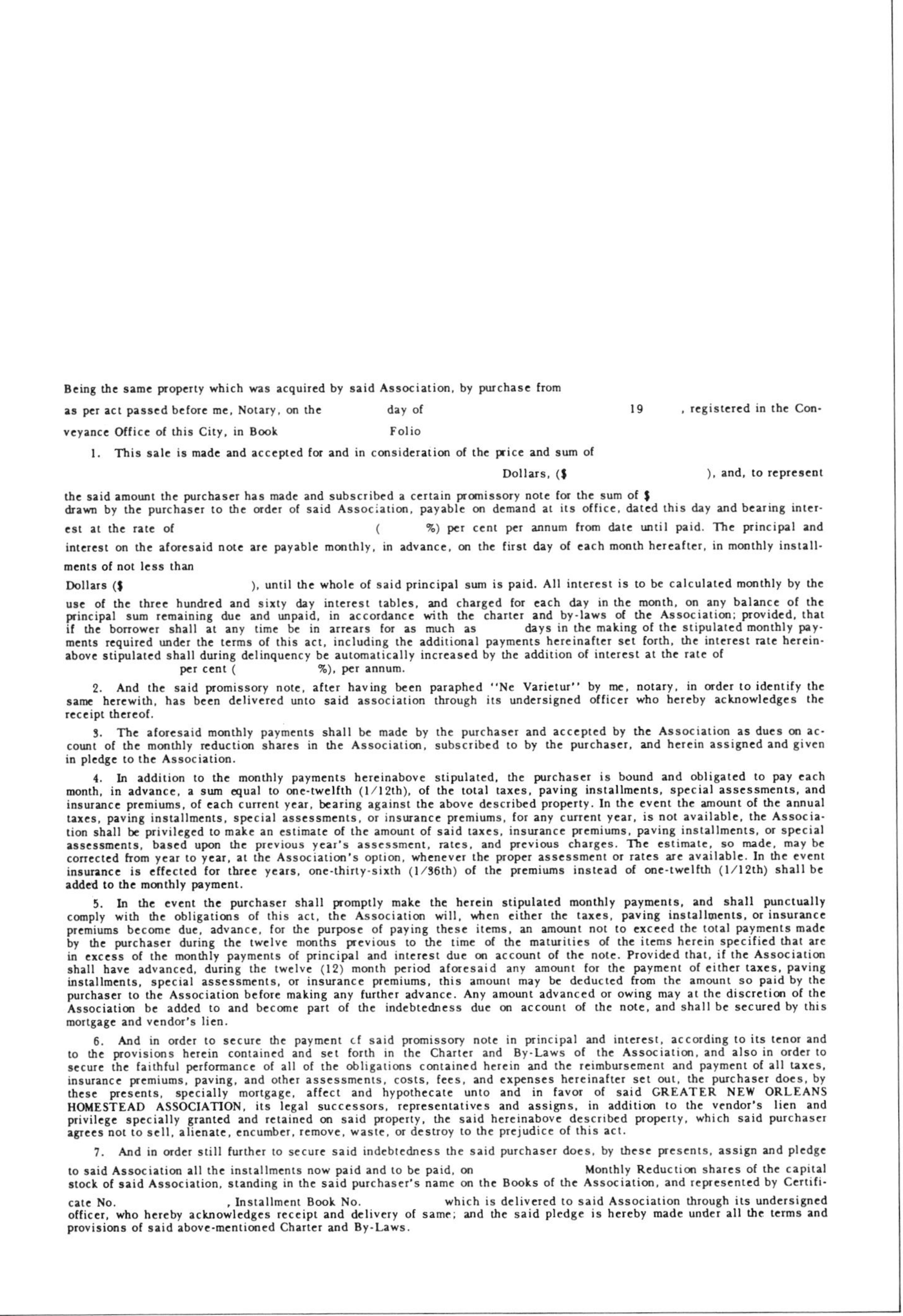

Being the same property which was acquired by said Association, by purchase from

as per act passed before me, Notary, on the day of 19 , registered in the Conveyance Office of this City, in Book Folio

1. This sale is made and accepted for and in consideration of the price and sum of Dollars, ($), and, to represent the said amount the purchaser has made and subscribed a certain promissory note for the sum of $ drawn by the purchaser to the order of said Association, payable on demand at its office, dated this day and bearing interest at the rate of (%) per cent per annum from date until paid. The principal and interest on the aforesaid note are payable monthly, in advance, on the first day of each month hereafter, in monthly installments of not less than Dollars ($), until the whole of said principal sum is paid. All interest is to be calculated monthly by the use of the three hundred and sixty day interest tables, and charged for each day in the month, on any balance of the principal sum remaining due and unpaid, in accordance with the charter and by-laws of the Association; provided, that if the borrower shall at any time be in arrears for as much as days in the making of the stipulated monthly payments required under the terms of this act, including the additional payments hereinafter set forth, the interest rate hereinabove stipulated shall during delinquency be automatically increased by the addition of interest at the rate of per cent (%), per annum.

2. And the said promissory note, after having been paraphed "Ne Varietur" by me, notary, in order to identify the same herewith, has been delivered unto said association through its undersigned officer who hereby acknowledges the receipt thereof.

3. The aforesaid monthly payments shall be made by the purchaser and accepted by the Association as dues on account of the monthly reduction shares in the Association, subscribed to by the purchaser, and herein assigned and given in pledge to the Association.

4. In addition to the monthly payments hereinabove stipulated, the purchaser is bound and obligated to pay each month, in advance, a sum equal to one-twelfth (1/12th), of the total taxes, paving installments, special assessments, and insurance premiums, of each current year, bearing against the above described property. In the event the amount of the annual taxes, paving installments, special assessments, or insurance premiums, for any current year, is not available, the Association shall be privileged to make an estimate of the amount of said taxes, insurance premiums, paving installments, or special assessments, based upon the previous year's assessment, rates, and previous charges. The estimate, so made, may be corrected from year to year, at the Association's option, whenever the proper assessment or rates are available. In the event insurance is effected for three years, one-thirty-sixth (1/36th) of the premiums instead of one-twelfth (1/12th) shall be added to the monthly payment.

5. In the event the purchaser shall promptly make the herein stipulated monthly payments, and shall punctually comply with the obligations of this act, the Association will, when either the taxes, paving installments, or insurance premiums become due, advance, for the purpose of paying these items, an amount not to exceed the total payments made by the purchaser during the twelve months previous to the time of the maturities of the items herein specified that are in excess of the monthly payments of principal and interest due on account of the note. Provided that, if the Association shall have advanced, during the twelve (12) month period aforesaid any amount for the payment of either taxes, paving installments, special assessments, or insurance premiums, this amount may be deducted from the amount so paid by the purchaser to the Association before making any further advance. Any amount advanced or owing may at the discretion of the Association be added to and become part of the indebtedness due on account of the note, and shall be secured by this mortgage and vendor's lien.

6. And in order to secure the payment of said promissory note in principal and interest, according to its tenor and to the provisions herein contained and set forth in the Charter and By-Laws of the Association, and also in order to secure the faithful performance of all of the obligations contained herein and the reimbursement and payment of all taxes, insurance premiums, paving, and other assessments, costs, fees, and expenses hereinafter set out, the purchaser does, by these presents, specially mortgage, affect and hypothecate unto and in favor of said GREATER NEW ORLEANS HOMESTEAD ASSOCIATION, its legal successors, representatives and assigns, in addition to the vendor's lien and privilege specially granted and retained on said property, the said hereinabove described property, which said purchaser agrees not to sell, alienate, encumber, remove, waste, or destroy to the prejudice of this act.

7. And in order still further to secure said indebtedness the said purchaser does, by these presents, assign and pledge to said Association all the installments now paid and to be paid, on Monthly Reduction shares of the capital stock of said Association, standing in the said purchaser's name on the Books of the Association, and represented by Certificate No. , Installment Book No. which is delivered to said Association through its undersigned officer, who hereby acknowledges receipt and delivery of same; and the said pledge is hereby made under all the terms and provisions of said above-mentioned Charter and By-Laws.

Figure 3-1 A sample mortgage form for the state of Louisiana.

METHOD OF APPLYING PAYMENTS:

8. The purchaser further covenants and agrees, and is specially bound and obligated as follows:

That the monthly payments, made and accepted as dues on the aforesaid shares, together with any and all additional payments, as herein specified, shall be added together, and the aggregate amount thereof shall be paid by the purchaser each month, in advance, in a single payment, and shall be applied by the Association in the following manner:

(a) To the payment of interest on the aforesaid note, at the rate hereinabove stipulated, on any balance of the principal remaining due and unpaid.

(b) To the re-payment of any amount which the Association shall have advanced for the payment of taxes, paving assessments, insurance premiums, special assessments, transfer fees, repairs, costs, attorney's and notarial fees, and expenses of any nature, together with interest thereon at the same rate as that charged on the principal sum.

(c) The balance, if any, to be applied to the reduction of the principal sum.

The above order of imputation may be changed at the discretion of the Association, or the remainder after payment of interest, may be credited in such manner as the Association may determine.

(d) Any deficiency in the amount of the aggregate monthly payment shall constitute a default under this act.

9. The balance due on the aforesaid note shall, at any time, be the amount of the face of said note, plus any amount which may have been advanced by or due to the Association, on account of taxes, insurance premiums, assessments, transfer fees, paving assessments, repairs, costs, attorney's and notarial fees, and/or expenses of any nature, together with interest thereon at the same rate as hereinabove stipulated, minus such sums as shall have been applied against the borrower's indebtedness.

10. That he will promptly pay all taxes, assessments, water rates, and other governmental or municipal charges, fines, or impositions, for which provision has not been made herein, and promptly deliver the official receipt therefor to the Association. If the purchaser fails to make such payments the Association is hereby authorized at its option to make them, and any sums so advanced shall bear interest at the same rate as, and become a part of, the principal debt from the date of payment, and shall be secured by the pledge and mortgage herein granted and by the vendor's lien herein retained.

11. That sums advanced by the Association for the payment of any taxes, special assessments, premiums of insurance, or any other charges against the property, or for any expense or cost whatsoever shall bear interest at the same rate as, and become a part of, the principal debt from the date of payment, and the reimbursement thereof shall be secured by the pledge and mortgage herein granted and the vendor's lien herein retained.

INSURANCE:

12. That he will insure and keep the building and improvements now existing, or hereafter erected, on said grounds, constantly insured, in such sums as the Association may require, against loss by fire, windstorm, tornado, and such other risks as the Association may hereafter require, in some good and solvent company or companies (which company or companies must be acceptable to the Association) until full and final payment of all the indebtedness hereunder, and will transfer and deliver the policy or policies of such insurance or insurances, and their renewals, to the Association, or assigns; in default of which the Association, or assigns, is hereby authorized, at its option, to avail itself of the rights hereinafter set forth, or to cause such insurance to be made and effected at the cost, charge, and expense of said purchaser.

13. That if the property covered hereby, or any pert thereof, shall be damaged or destroyed by fire or other hazard against which insurance is held, the amounts due by any insurance company shall, to the extent of the indebtedness then remaining unpaid, be paid to the Association, and, when so paid, may, at its option, be applied to the debt or be used for the repairing or rebuilding of the said property.

14. The purchaser also agrees in the event of loss, not to make settlement with any Insurance Company without the approval of the Association and hereby employs its Building Expert to adjust any loss, the fee to be paid by the Insurance Company, if such fee is provided for in the policy. If the policy does not provide for such fee, then the purchaser shall pay said fee.

REPAIRS:

15. The purchaser agrees to keep the property in good repair; that failure to do so within ten days after written notification by the Association to the purchaser, at his last known address, shall ipso facto mature the entire balance due said Association by the purchaser and shall authorize executory process by the Association under the general terms and provisions relating to executory process as herein provided; or if the Association prefers, the purchaser hereby authorizes said Association without putting in default to cause necessary repairs to be made at the purchaser's expense, and charge same plus the cost of supervision to the purchaser's obligation.

REIMBURSEMENT FOR ADVANCES:

16. The said purchaser hereby authorizes said Association to reimburse itself from any credits or pledges, including paid up stock or optional payment shares, in its possession and belonging to said purchaser, for all payments due it and for all payments or advances made by it by reason of non-payment by said purchaser of any obligations due hereunder, including attorney's and notary's fees, and costs.

ASSIGNMENT OF RENTS:

17. In the event of violation of any of the terms, conditions, or covenants of this act, such violation shall ipso facto effect an assignment of the leases and rents to said Association and the purchaser hereby authorizes the Association to collect, by legal process, if necessary, rentals due by tenants of the purchaser occupying the property hereinabove described and to have delinquent tenants evicted. After deducting the cost and/or eviction, the net proceeds shall be credited to the purchaser, and the purchaser hereby specially agrees that when the account is delinquent, payments thereon may be credited in the manner the Association deems best.

CONDITIONS MATURING NOTE:

18. That in the event the purchaser should violate any of the conditions of this act, or should fail promptly to perform any obligation hereunder or be in arrears for a period of 13 weeks even if the purchaser shall have made prior payments in excess of the amount required, or upon the happening of any one or more of the events or conditions listed below, the entire balance due by the purchaser to the Association shall ipso facto, without demand, notice, or putting in default, immediately become due, exigible and payable, together with interest, taxes, premiums of insurance, attorney's and notarial fees, advances and all costs, expenses and other charges, whatsoever, the said purchaser hereby confessing judgment in favor of said Association and its assigns, for the full amount of said promissory note together with all interest, taxes, premiums of insurance, attorney's and notarial fees, advances, and all costs, expenses, and other charges whatsoever.

(a) Insolvency of the purchaser, application by the purchaser to be adjudicated a bankrupt, or the institution of involuntary bankruptcy proceedings against him, or the institution by or against the purchaser of any proceedings for the appointment of a receiver or syndic, or the filing of any proceeding seeking reorganization if the purchaser be a corporation; the purchaser hereby agreeing that the happening of any one or more of the incidents herein enumerated will not in any way enjoin, hinder or delay any legal proceedings which the Association has the right to institute under the provisions of this act; and the purchaser further agrees that no Order of any Court in any of the above enumerated proceedings will be sought to enjoin, hinder or delay the proceedings which the Association has the right to institute under the provisions of this act.

(b) The recordation or registry of any lien or claim, or the institution of any legal proceedings to enforce any lien or claim against the property; or if the property be seized or levied upon by an officer of court.

(c) Death or dissolution of the purchaser.

(d) The cancellation of any insurance covering the property, for whatever reason, if the purchaser fail immediately to replace said insurance in a company or companies satisfactory to the Association; or upon the inability of the purchaser to procure insurance protection required under the terms of this act;

(e) The use of the property for any unlawful purpose;

(f) The making of any repairs, additions, or alterations to the buildings or improvements on the ground herein conveyed, or allowing of any work to be done whereby any lien or privilege could result against the property, or in case of the actual or threatened alteration, repair or addition to, demolition or removal of any building on the property herein mortgaged, without previously obtaining the written consent of the Association;

(g) The sale or transfer of the property without the written consent of the Association.

19. The purchaser acknowledges and agrees that all of the provisions of this deed are of its essence; that any violation of any of the terms hereof; shall ipso facto mature the obligation of the purchaser and shall authorize said Associa ion to proceed by executory or any other process in accordance with the terms of this deed governing the procedure when executory or other process may be resorted to; and that the Association would not have made the present sale to the purchaser unless all of said provisions were agreed to by the purchaser.

Figure 3-1 A sample mortgage form for the state of Louisiana.

20. That in the event of any default or the violation of the conditions of this act, or the happening of any one or more of the events hereinabove mentioned, the Association shall have the right, without the necessity of demand or of putting in default, to cause the property herein described, together with all the improvements thereon, to be seized and sold under executory or other process issued by any competent court, or may proceed to the enforcement of its rights in any other manner provided by law, and the property may be sold with or without appraisement, at the option of the Association, to the highest bidder for cash, the purchaser hereby waiving the benefit of all laws relative to the appraisement of property seized and sold under executory or other process.

21. In the event of sale of said property under executory or other legal process said Association shall have the option of causing said property to be offered and sold by the Sheriff at the Sheriff's Sale, in parcels or as a whole.

22. That if legal proceedings are instituted for the recovery of any amount due hereunder, or if any claim hereunder is placed in the hands of an attorney for collection, the purchaser agrees to pay the fees of the attorney at law employed for that purpose, and such fees are hereby fixed at per centum (%) of the amount due, the minimum attorney's fee to be twenty-five dollars.

23. The Association may, at any time, without notice, release all or a part of the mortgaged premises from the lien or effects of this mortgage, grant an extension, or deferment of the time of payment for any indebtedness secured hereby, or release from liability any one or more parties that have become liable for the payment of said indebtedness secured hereby, without affecting the personal liability of the purchaser, or any other party liable, or hereafter becoming liable, for the payment of any of the indebtedness secured by this mortgage.

24. Upon application by the purchaser, or any future owner of the property hereinabove described, to the Association, for a reduction in the amount of the monthly payments, the Association shall have the right, through any of its designated officers, or committee, as its Board of Directors may authorize, to act on such matters. In case of such reduction, the certificate of any officer of the Association, under its seal, shall at all times, be conclusive and authentic evidence of the amount of the required monthly payment and of its having been fixed by competent authority.

25. It is hereby agreed and understood by and between the parties hereto that if the said purchaser shall make the herein stipulated payments promptly and punctually, then and in that case, the principal of said note shall not become exigible until the value of said shares of stock shall become equal to the amount of said indebtedness with all interest and costs that may be due upon the same, at the happening of which event, said stock and said indebtedness shall cancel each other, the stock and indebtedness being alike extenguished and thereupon it shall be the duty of any officer of said Association, to cancel said stock and give to said purchaser full receipt and acquittance, as well as to surrender to said purchaser the said abovementioned promissory note.

26. If at any time before the value of the purchaser's stock shall be equal to the purchaser's indebtedness, the said purchaser should desire to pay, settle, and cancel said indebtedness, said purchaser shall have the right to surrender to the Association in part payment of such indebtedness at their cash value, the shares herein pledged in this Association provided said purchaser pays in cash the balance of such indebtedness and, then it shall be the duty of any officer of said Association to cancel said stock and to give and render to said purchaser a proper receipt and acquittance as well as to surrender said note to said purchaser.

27. To have and to hold the said property and appurtenances to the said purchaser, said purchaser's heirs and assigns to their proper use and behoof forever.

28. The Association declares that this sale is made without warranty, not even for the return of the purchase price; but, the said Association subrogates the said purchaser to all the rights and actions of warranty and otherwise, which the said Association has or may have against all former owners of said property, fully authorizing the said purchaser to exercise the said rights and actions in the same manner as said Association itself might or could have done, the whole without recourse.

29. And the said Association declares that the fact that it does not warrant the title of said property is not to be construed as casting a doubt or cloud upon the title, for the reason that the property is vested in said Association temporarily only, and merely for the purpose of obtaining the security of a vendor's lien mortgage in accordance with the special provisions of the law of Louisiana governing Building and Loan Associations.

30. This sale and mortgage is made to the purchaser in consideration of his membership in the Association and the member-purchaser agrees that the Charter and By-Laws of said Association are and shall be binding upon him and do by reference form a part of this contract, and he further agrees that any amendment to the said Charter or By-Laws of the said Association, properly and regularly adopted, shall be likewise binding upon him and so long as any part of the obligation hereunder created shall be unpaid, said amendments to the Charter and By-Laws shall become and form a part of this contract to the same extent as if the same were written in extenso herein.

31. The purchaser agrees that the failure of the Association to exercise any of its rights, privileges, or options, at any time, is an indulgence granted to him and shall not constitute a waiver of its rights to exercise the same at any other time. Nothing, in this act contained, shall be construed so as to limit any right or remedy granted or available to the Association under any provision of law, or its charter and by-laws.

32. The covenants herein contained shall bind, and the benefits and advantages shall inure to, the respective heirs, executors, administrators, successors, and assigns of the parties hereto. Whenever used, the singular number shall include the plural, the plural the singular, and the use of any gender shall be applicable to all genders.

33. It is agreed that the right herein given to said Association to pay taxes, effect insurance, collect rents, make repairs, or any other right granted herein, shall not be construed as obligating said Association to do any or all of these things, or making it liable in any manner for its not doing any or all of them.

34. The production of Mortgage, Conveyance, and United States District and Circuit Court Certificates is hereby waived by the parties hereto, who relieve and release me, Notary, from all responsibility and liability in the premises for such nonproduction.

35. The purchaser waives and renounces in favor of the vendor, and any future holder of said note, any and all homestead rights in and to the above property which he has or may have under the Constitution and laws of the State of Louisiana.

36. Any paving or other lien or assessment chargeable against the herein conveyed property is assumed by the purchaser. All the taxes on the said property, up to and including those due and exigible in the year 19 , are paid, as will appear by reference to the vendor's act of purchase hereinabove mentioned. The taxes for 19 , are assumed by the present purchaser.

THUS DONE AND PASSED at my office in the City of New Orleans, on the day, month and year first above written, in the presence of the undersigned competent witnesses, who have hereunto signed their names with the appearers and me, Notary, after reading of the whole.

GREATER NEW ORLEANS HOMESTEAD ASSOCIATION

WITNESSES:

By:..

President.

Figure 3-1 A sample mortgage form for the state of Louisiana.

It becomes apparent that if you are able to pay off your mortgage rapidly, it cuts your charges markedly. The interest on a five-year loan at 7 percent amounts to $7,523.00 in contrast to $55,806.80 at the same rate for 30 years –a difference of $48,283.80. A 20-year loan at 8 percent costs $25,364.40 less than the same loan for 30 years.

Interest rates in the 1973-1974 era varied from 7 to 10 percent. You can rest assured that there will always be variations in this highly competitive area of home loans. Therefore, it is wise to shop around and check the different types of institutions–banks, savings and loan associations, mortgage companies, and insurance firms.

You will have a stronger bargaining position if you can make a larger down payment. Lending institutions are concerned about the safety factor, and a larger down payment (other things being equal) reduces their risk. It is true that the mortgagee (person borrowing the money) can be taken to court by the mortgagor (lender) for failure to meet payments, but legal proceedings are costly and time consuming. All lenders wish to avoid this action.

I normally recommend a 20 percent down payment–more if feasible.[1] Note the difference in one neighborhood when two individuals bought similar houses for $40,000.

1 Percent down	2 Amount of down payment	3 Principal amount of loan	4 Duration of loan (years)	5 Interest rate	6 Total interest	7 Total cost (2+3+6)
20	$8,000	$32,000	30	7½	$48,550	$88,550
5	$2,000	$38,000	30	7½	$57,656	$97,656

Lenders are anxious to provide funds to "blue chip" borrowers–individuals who make a healthy down payment and have an excellent credit rating. They can be expected to meet all their loan obligations and so receive the best terms. Borrowers should strive for this category because they will be eligible for the lowest interest charges and thus will place themselves in the strongest bargaining position. Although a good credit risk may wish to make the maximum down payment and pay off his loan as rapidly as possible, the loan officer might try to have him take the mortgage for a longer time. The lender's rea-

[1] If you build a house knowing you will leave the area within three years, you can support a case for making the minimum down payment and extending the loan over the longest term. The reason given is that more people will be able to take over the loan.

soning is apparent from the statistics which show that home costs increase in direct relation to the duration of the mortgage. He may try to convince the borrower that the actual interest charged does not tell the whole story.

The loan official may mention the income tax write-off as a great advantage. I don't agree with this reasoning at all. If someone is in the 30 percent tax bracket (on the average) for the life of the loan and it costs 8 percent to borrow money, it means the tax deduction may reduce it as low as 5.4 percent.[2] But that is still a sizable charge and a person would be fortunate to do as well (after taxes) on his own investment portfolio. My recommendation: forget the tax advantage and proceed to use the money saved on interest to build your own sound money management program.[3]

Types of Mortgage Loans

Next let us turn to the three types of mortgage loans available—FHA, VA, and conventional.

GI Loan[4]

Since the end of World War II, more than 9 million individuals have purchased homes with the assistance of GI loans under the guaranteed home-loan program of the Veterans Administration (VA).

The present rate of interest on a GI loan is 8½ percent. The VA will guarantee no more than 60 percent of a loan, the top portion of the loan, in an amount not to exceed $12,500. Eligible individuals first apply to a lending agency. The major firms doing this type of business are mortgage brokers.

The GI loan provides the following advantages:

1. The interest rate is normally lower than the going rate at the time. For example, the VA charges 8½ percent, whereas in this area conventional loans are running between 9¼ to 9½ percent.

2. With a VA guaranteed loan, you have the right to repay your loan at any time without penalty.

[2] I say *may* because taxpayers receive up to $2,000 as an automatic deduction. If an individual has write-offs of only $1,000 then his taxable savings from interest is reduced by this amount.

[3] For a detailed discussion of this topic see the author's *Guide to Personal Finance: A Lifetime Program of Money Management*, Prentice-Hall, 1972.

[4] The phrase "GI loan" means a loan made by a private lending institution to a veteran for any eligible purpose, pursuant to Title 38, United States Code. Source: *Loans For Veterans*, VA Pamphlet 26-4, revised May 1971, p. 2.

3. No down payment is required.
4. The loan may be made for up to 30 years.

The disadvantages you may encounter are these:

1. There will undoubtedly be red tape when dealing with the government.
2. Add-on points[5] are often charged that may absorb much of the savings gained from the lower interest rate. Although the VA will not permit point payments legally, there are ways to do this "under the table."

FHA

The Federal Housing Administration (FHA) also provides an opportunity to obtain a loan at less than the going rate. The FHA, like the VA, guarantees a loan for the lending agency. The procedure is similar, in that you go to a lending agency first to get its approval for a loan. Anyone who has good credit, steady work, and the necessary cash down payment is eligible for an FHA loan. At the present time, the maximum mortgage that the FHA will guarantee for a single-family dwelling is $33,000.

The FHA calculates the maximum insurable loan by taking:

97 percent of the first $15,000
90 percent of the next $10,000
80 percent over $25,000

The advantages of the FHA loan are as follows:

1. There is a smaller down payment than with a regular loan. If an individual purchases a $15,000 home, his down payment is only $450. (But the buyer may have to pay closing costs similar to those in a conventional loan.)
2. The interest rate is normally more reasonable. It is currently 8½ percent plus ½ percent FHA mortgage insurance.
3. The period of payment may be longer, with a 30-year maximum.

The prime disadvantage that I found, from talking with people who have gone the FHA route, is the amount of time consumed in bureaucratic delay.

Conventional

The key advantage to the conventional loan is that it eliminates the governmental red tape. However, it is often more costly; and also, in periods of high

[5] A one time charge based on the amount of the mortgage, which may be utilized by lending agencies. One point is equal to a one percent additional cost for one year. For example, let us assume a homeowner is required to pay 4 points above the 7% loan on his $40,000 mortgage. Such a charge would result in his paying an additional $1600.

interest rates, the lending agencies are more selective and these loans may be difficult to obtain.

In the 1973-1974 capital market, it was generally not possible to secure the best deal on a conventional loan. Two examples illustrate the picture in one area: First, the branch manager of a large savings and loan firm told me he was making no GI or FHA loans—money was too scarce and there were better opportunities in the regular home loan. He was making conventional loans up to 80 percent of appraised value. On a $20,000 home, this meant a $16,000 maximum loan, with interest charge of 7¾ percent and monthly payments to be completed in 20 years. So a buyer had to have $4,000 of his own money available or secure a second mortgage from some other source—at a higher rate of interest.

In the second case, a prominent lending institution on the West Coast was requiring a 25 percent down payment on a $25,000 house. Although the bank would loan $18,750 at 8 percent, they stipulated that if the prime interest rate was raised before the contract was closed, the borrower had to pay the higher price. In addition, they would not grant the loan for a period in excess of 20 years. Consequently, monthly payments, including interest, taxes, insurance, and so on, approximated $260. Other costs of closing the deal amounted to an added $1,000 (tax reserve, title company fee, recording fee, and so on). In this situation, the buyer had to muster at least $7,250 to obtain the $25,000 house.

What are current conditions with respect to the three categories of loans? In view of the red tape, some savings and loan associations will not utilize FHA. One savings and loan official was asked why he took VA. He commented:

> There is governmental interaction but it isn't as bad as the FHA. Mortgage companies handle FHA. Although you have to put down only 5 percent with FHA, mortgage companies have other means of making money. Some ask a commitment fee. This means that if you indicate your intentions in taking out a mortgage, you are charged for the paperwork involved even if you don't go through with the deal. There are also discount points which may run from 6 to 8 percent the first year. Thus, if your house and lot are valued at $30,000 you would actually spend $1,800 to $2,400 more for the initial purchase by paying 6 to 8 discount points. It is very important for you as a homebuilder to know exactly what all costs are in making a loan.

Computing All Costs

All types of lending agencies try to maximize their costs. It behooves each borrower to be sure he has calculated all charges correctly. With interest rates high, it pays to make the maximum down payment. Why let the bank charge you 8 percent when your money only earned 5 percent? The question of points should also be given a hard look. The number of points charged in my area ranged from two to eight. On $20,000, this one-time charge would amount to $1,600 at the highest rate. Another "hidden" cost is supervision; that varied from zero to one percent. I observed the amount of supervision given my neighbor's home—about 30 minutes of casual observation during the entire construction period. I can think of better ways to spend $200. A performance bond or placement in escrow also involves a one percent charge.

What about insurance and taxes? The lending agencies love to take care of these items for you. A doctor friend told me his bank representative advised him that the bank could get a better tax assessment, "so let us handle it." The doctor commented:

> My family has lived in this area for four generations; we know all the tax angles and can get the best deal. What the institution wanted was for me to pay my taxes up to a year in advance. They would invest my money until payment was due—a great deal for the bank, but I was getting taken. On insurance, I found their rates higher than the company we use for my medical affairs, so I told them impolitely to go to hell.

Cut your borrowing costs to the minimum. It makes good sense to pay your own taxes and insurance. Shop around for insurance and obtain the policy best suited to your needs. During construction, obtain a policy protecting you from risks such as injury to workmen and others,[6] theft, destruction of materials, and damage to neighbors' property. Such coverage can be obtained locally.

Convincing Your Lending Agency

I have pointed out the importance of having your records available when you sit down with the lending agency, particularly when you subcontract your house. Your job will be to convince the loan officer that you can build a satisfactory home. He may point out that you lack experience to do the job properly and have inadequate buying power to obtain the best prices from subcontractors and supply houses. He may also indicate that interim interest

[6] Subs often have this insurance; it is important to obtain a copy of it before they begin work. The "Certificate of Insurance" should be from a reliable company. It itemizes type of insurance (employer's liability, workmen's compensation, comprehensive general), policy number, expiration date, and limits of liability.

between the time you start building and moving in could be considerable. For example, if it should take one year to build a home and you borrowed $30,000 (average amount) at 8 percent, it would cost $2,400.

The loan officer's points are valid in some cases. But you should emphasize your research and knowledge in the homebuilding field as well as the fact that your subcontractors will be highly qualified. It is also helpful to indicate your close (daily) supervision.

If you have done your homework, it will be apparent to the loan officer. You may have found a first class carpenter who will give you a "turn-key" job. That is, he will be present from the initial survey until you move in. Another approach is to have a relative or friend who is in the contracting business available to help you. Some people who build their own homes have engineering backgrounds or are successful managers. One loan officer put it this way: "I want to be damn sure that when we loan our money, the man will be capable of meeting his payments. If he doesn't come up with an adequate home, or interim interest eats him up, I could be in trouble."

A normal agreement between you and a savings and loan association is for the institution to hold in escrow the entire money needed for the house. If the house will cost $30,000 to build, this amount will be set aside and payments made as subcontractors submit their invoices. You only pay interest on the amount used and it is at the standard going rate.[7] It is desirable to talk with at least three reputable lending institutions in order to get the best deal. A Dun and Bradstreet report can be obtained through your bank, and respected leaders in the community can also give you good advice. Friends and relatives who have used these lending agencies for home buying over the years can provide valuable information.

Be sure to obtain every cost involved, and find out whether the rate charged will fluctuate over the years, based on the changing interest rates. Ideally, you should try to secure a loan where the institution will reduce interest in a declining period and not increase it again when the rate rises. In this period of high interest, it is very important for you to have an escape clause. Keep in mind that sometime during the duration of long-term loans (20 to 30 years) interest rates will decline. You must have the opportunity to refinance at the lower rate or pay off the loan—*with no penalty charge for*

[7] Make certain you have a written agreement with the lender that you will be charged interest only on the amount actually used. There have been cases where borrowers paid interest on the entire amount from the day it was placed in escrow.

early payment. Be sure to read all the fine print. If you don't understand something, ask the loan officer to explain. If still in doubt, check with an authority—preferably a respected lawyer qualified in this field.

What about buying your lot prior to building your home? Savings and loan associations will lend you dollars on a lot if you intend to build in a couple of years. These institutions will not make such loans for speculation purposes.[8] The rate is higher than for a regular mortgage loan. If questioned on the interest charged for a lot, some savings and loan officials may use the term 6½ percent add on. This in fact means 11½ to 12 percent. I recommend that you buy your lot just before you build. You obtain a lower interest rate on your lot in a package deal and you don't tie up your money in idle land for a long period of time. Besides you may change your mind about building in that particular location and have a difficult time selling it quickly at a fair price. This is particularly true in our present high-interest era.

Banks For Interim Financing

If you should seek out a commercial bank for financial assistance, it will normally provide only interim financing. It may, however, work with a savings and loan association which will take over the mortgage. As a result, you may only need to utilize one title search, lawyer, and so on. My neighbor used this dual method to subcontract his house. He did a beautiful job building his home and still ran his small business successfully. He commented:

> I can't over-emphasize the importance of having an officer at the bank who knows you and believes in you. What happens is that this "friend in court" participates in the daily meetings that review all loan requests. Here the decisions are made to loan or not to loan. This man can be very influential if he supports your position, pointing out that you are a good risk and capable of doing the subcontracting. But to gain his support initially, I had to prepare a detailed budget and other data as well as secure a qualified carpenter who would help supervise the job.

This friendly relationship warrants a word of caution, particularly if you plan to use this friend in the future. I recently heard of a bank that got burned by loaning a good friend $100,000 to subcontract his house. He proceeded to spend money like it was going out of style. No valid records were kept and he had no written agreements with his subcontractors. When the

[8] There is considerable variation in the philosophy of savings and loan associations throughout the country. For example, Texas has a liberal policy. Currently, it is possible to borrow up to 80 percent of appraised value on raw land.

money ran out, he came back to the bank for an additional $50,000. This time his good friend said "No." An investigation discovered that he had overspent his original estimate by $70,000 and the value of his property totaled only $35,000. The bank foreclosed on him.

Insurance Company Loan

It may be to your advantage to contact local representatives of highly respected life insurance companies. If you go this route, they will require you to take out a life policy for the amount of the loan. This makes a nice profit for them annually on both the policy and the mortgage interest. You may find, however, that their interest costs are lower. If this is true and you have a need for life insurance, this could be your best deal. The younger homebuilders, in particular, might be interested in view of the lower premiums for their age bracket. Another advantage of obtaining a mortgage from some insurance companies (mutuals) is that the annual dividends can be applied to the loan and thus reduce the length of time required to pay off the mortgage.

A Second Mortgage

You may have your heart set on building your own home immediately. This decision might be made with full knowledge of the risks involved, including high interest rates. Let's assume your first mortgage would not cover all your financial needs. The agency from which you obtain the first mortgage would hold your home and land as collateral. You estimate your building costs at $40,000, and you have $4,000 in cash and a savings and loan association agrees to loan you $30,000. You are now short $6,000; one way to secure this money is to obtain a second mortgage.

The company willing to lend the money on a second mortgage will demand a higher rate of interest. If you failed to meet your payments, you might be forced to sell your property, and if the sale brought only $25,000, it could all go to the first mortgage holder.

The second mortgage holder would only obtain his money after the first mortgage is completely paid off. In order to protect himself, he will not only charge a higher interest rate but he will also lend you the money for a shorter period—usually no longer than 10 years.

If you decide to take out a second mortgage, it is important to go to a reliable firm and to find out all costs involved. For example, the interest charge on the $6,000 may be within the legal limit of your state (10 percent for example), but there also may be hidden charges. Unfortunately, there is no consumer protection against these expenses. Such costs might even include a placement fee for the company who leads you to the second mortgage broker. Read the second mortgage agreement carefully. *Better yet, avoid taking out a second mortgage.* If necessary, borrow from a friend, relative, your insurance policy, or take out a personal loan. Ideally, it is best if you can subcontract your home without having to borrow a penny. This precludes any delay due to obtaining the loan and securing clearance on payment of invoices. It also eliminates brokers' snooping into your personal financial affairs. Lending agencies can be difficult at times when you need their financial support the most.

If you can save sufficient funds for a smaller home, it may be preferable to borrowing for the expensive house. The initial subcontracting gives you invaluable experience that can result in a more successful job the next time.

Summary

Adequate money is an essential ingredient in successfully subcontracting your home. Consider banks, savings and loan associations, insurance companies, and mortgage firms for the best deal. Be familiar with the pros and cons in utilizing the VA, FHA, or conventional loan.

I recommend a minimum of 20 percent down payment and paying off your mortgage as rapidly as possible. It is so important to consider all money costs in homeownership. Initially you must obtain sufficient funds to buy your lot and meet all payments to your lending agency, subcontractors, and suppliers. Once settled in your home, there are mortgage payments, taxes, insurance, heat and utilities, maintenance, and major expenditures for large appliances and furniture. If there are second mortgage payments to be met, there is the added risk of foreclosure proceedings and of losing the house you worked so hard to build.

In summary, I must emphasize the importance of having all the facts available prior to conferring with your loan officer. The more thorough your financial data, the better chance of obtaining the loan and meeting your payments in years to follow.

CHAPTER 4
Building Your Home

Okay, the money is available and your planning is complete. You're now ready to build your home. As mentioned in Chapter 1, the functions of management are threefold: (1) plan what you are going to do, (2) do what you planned, and (3) check on what you have done. (See Figure 1-1.) To perform these managerial functions, you need men, money, and materials. The key to coordinating it all rests with you as the "boss," or manager, who must make sound decisions.

You are now ready to carry out your building plans. It is helpful to recheck your survey. Be certain you are building on a lot you own. Also review local requirements to be sure you build in accordance with city and subdivision restrictions.

At this point, you are ready to contact that first-rate carpenter you have selected. He can help you immeasurably. As stated previously, you pay more if he gives you a "turn-key job," meaning that he is available from the time he lays out the forms for the foundation until the house is completed and the key is turned over to you. Mentioning this experienced carpenter to your lending agency, I am confident, helps in obtaining the loan.

There is no substitute for the experience of a competent carpenter. Most of my subs had little education but had a great deal of common sense, loyalty, and pride in their workmanship. The experience convinced me that our current teachings in personnel and management are worthwhile and that human understanding, motivation, self-actualization, coordination, communication, and incentives really work. At times, I had six or seven different subs working together—carpenter, heating and air conditioning expert, plumber, painter, roofer, sheet metal man, and pool installer—and if there hadn't been any cooperation among them, there could have been trouble.

Your Job Responsibilities

You can avoid possible conflict by having only one subcontractor on the job at a time; but time is money. Besides, more losses occur (theft and pilferage) and prices rise in the interim. It seems as if delay builds on delay. Therefore, your big responsibility is to be on the job. I worked approximately 12 hours a day, including time on the site, taking care of financial records, doing the buying, and coordinating the day-to-day activity. Most people can't spare this block of time but do utilize vacations, weekends, and evenings. A friend who was successful in this venture had his wife perform the supervision; another hired someone to oversee the project when he wasn't available.

I participated in the building wherever feasible—assisted in installation of heating and air conditioning, hauled and sorted bricks, and provided service to meet emergency needs.

Prompt decisions are required daily; and if you are not present, the sub may walk off the job. After all, he is paying his people by the hour. If you don't have the necessary materials or if you don't tell him what's expected, why shouldn't he leave? And it may be a long time before he returns—another job may now take priority.

Your presence on the job also prevents mistakes. For example, if you are paying the carpenter for paneling and the insulation has not been completed when he does this, you may never find out why it costs you so much to heat or cool the house.

We normally had no workmen on the job on weekends. This permitted my wife and me to shop for materials, clean the area, and bring records up to date. Whenever possible, we conducted business on weekends at our housing site. It saved time to have representatives visit us with samples and catalogs of rugs, drapes, lighting fixtures, and so on. We also found that the subs in New Orleans love the weekends for fishing and other entertainment. It cuts costs and increases productivity to provide this time off. Most important, subs usually charge time and a half (sometimes double time) for Saturday and Sunday work.

Key Points

You should initially prepare a step-by-step table, showing the sequence of events, comparable to the one we used in building our home (see Figure 4-1). A PERT chart can also be very useful during this construction phase (see Figure 4-2). If you do prepare a PERT chart and timetable, *they must be made to suit your requirements.*

1. Begin building phase. Discuss techniques to meet needs—PERT, management model, organizational and control steps.
2. Lay stakes for piling.
3. Piling driven.
4. Temporary electrical hookup completed.
5. First city inspection completed.
6. Lumber delivered.
7. Exterior forms laid out.
8. Foundation plumbing completed.
9. First plumbing inspection.
10. Grading and preparation completed.
11. Lumber and other supplies delivered.
12. Termite inspection completed.
13. Concrete poured for foundation.
14. Garage secured.
15. Frames completed by carpenter.
16. Sheathing for exterior completed.
17. Roof covered with waterproof material.
18. Roof completed.
19. Windows and glass doors installed.
20. Inside plumbing completed.
21. Air conditioning and heating completed.
22. Bricks for home delivered.
23. Electric turbine vent completed.
24. Electrical and phone wiring completed.
25. Iron set above windows.
26. Insulation for walls and attic completed.
27. Bricking complete.
28. Sheetrock® complete.
29. Slate porch complete.
30. Gutters installed.
31. Terrazzo floor complete.
32. Concrete for drive and walks complete.
33. Kitchen cabinets and plumbing fixtures installed.
34. Tile completed.
35. Outside air conditioning completed.
36. Inside doors installed.
37. Ground cleaned and prepared for landscaping.
38. Painting completed.
39. Fence completed.
40. Rugs installed.
41. Flower and plant landscaping completed.
42. Plans filed.
43. Final inspection by city.
44. Light fixtures completed.
45. Wallpaper completed.
46. Drapes installed.
47. Insurance converted from construction to home.
48. Job completed. Move in.

Figure 4-1 A step-by-step approach of the sequence of events in the building phase.

Building Your Home

What are some of the key points to look for when you are actually building your home? Our PERT chart helped us complete our home in two months, in contrast to our neighbors who averaged six months for comparable homes. Examine the items listed in Figure 4-1. Later in the chapter, we will look at photographs which highlight other important features. It may be helpful to refer to Appendix I, showing the blueprints of our home.

Start of Building Phase

At this time, you should prepare your PERT chart and timetable. The same format that you developed at the beginning of the planning phase can be applied. You must realize that these are guidelines only and should be modified as you proceed with construction. Figure 4-2 shows our PERT chart representing the optimistic time for completion of the project. You may wish to develop as many as three PERT charts—optimistic, average, and pessimistic. I suggest you make a few copies; keep one at home and have at least two others for you and your wife to carry. One of my friends posted the master chart in his bedroom. He would check it each evening and make modifications where required. When an activity was completed, he would mark it off.

At the beginning of this building phase, you should review the material presented in the first three chapters. Familiarize yourself with the management guidelines and the model (Figure 1-1).

Foundation

A good foundation is essential in the building of your home. Be present during the excavation and check to see that the materials used meet your specifications. Many homes are constructed with basements. The objective here is to have a basement that will be free of water which might penetrate through the floor and walls. Make certain that there is good outside grading that will carry any water away from the house.

In our case, we could not build a basement—New Orleans is below sea level. In order to prevent the house from sinking, pilings were required. If you use pilings, be sure that your carpenter or surveyor places the stakes in accordance with the architectural drawings. It is desirable to have the pilings driven in the ground shortly after the markings have been placed. Some youngsters delight in rearranging these stakes; this could mean inadequate support. It also pays to watch the pile-driving crew to make certain they put the pilings precisely where the stakes are placed.

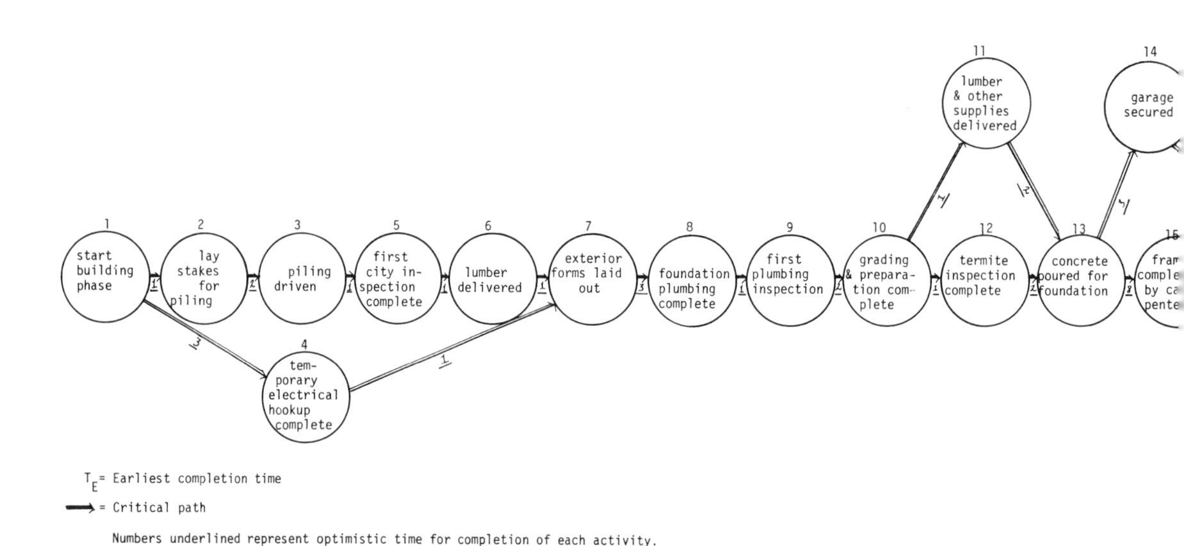

Figure 4-2 PERT chart used during the building phase of the Stillman house.

Be present when the pilings are delivered to see that they are not dumped in the street or on the neighbor's property. This also provides an opportunity to see that they measure up to the quality ordered. You will be charged for the number utilized; so when the job is done, make your own independent count. The less blows it takes to drive them in the ground, the quicker the job can be done. However, if you have neighbors with valuable china or other breakables, it would be courteous to notify them when this work will take place. You should also caution the company and suggest they take more time in sinking their pilings. I observed only one operator who did a sloppy job. The contractor had specified 35-foot pilings of adequate diameter but his workmanship resulted in several breaking off at 20 to 25 feet. If the ground is hard, the pilings can be driven only to "refusal." This may result in their being able to penetrate the ground to perhaps only 20 to 30 feet.

Electrical Hookup

Call the electrical company to install a temporary hookup while the piles are driven. Your carpenter will need this power to cut the lumber for the

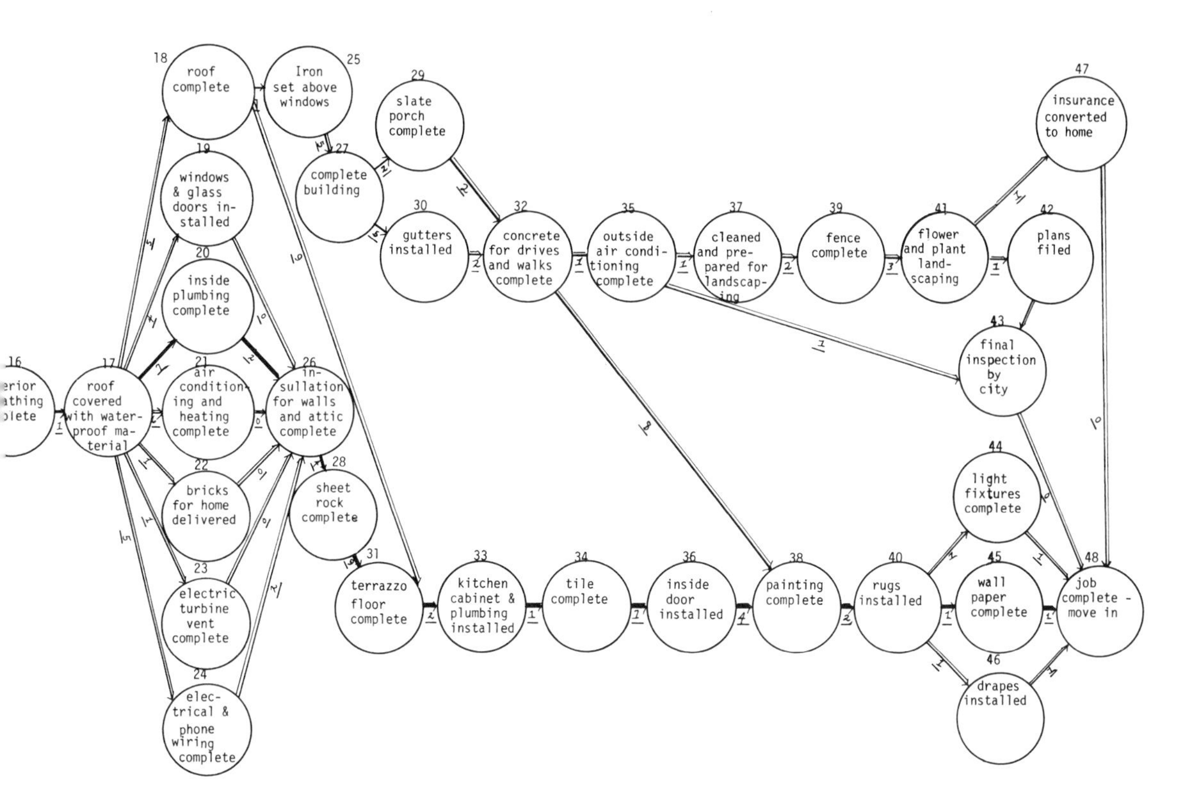

exterior forms. Although it is possible for him to use a gas-powered saw, this is expensive and not as efficient. You could ask a neighbor for permission to hook into one of his outlets as a temporary expedient. If this is necessary, I suggest you offer to pay him and thereafter maintain his friendship. Your subcontractors may also use his water and park in front of his premises. You will be living beside him soon and it helps to start off on friendly terms.

First City Inspection

While the pilings are being driven, it is time for the city inspector to determine if they are satisfactory. It is incumbent upon the pile-driving company, or contractor, to inform the city at least 24 hours before work is to begin. You and the lending agency representative should be present when the inspector arrives. Be nice to city inspectors; they can offer helpful advice. I talked with a city inspector to verify his procedure:

> I consider the piling check the most important one. If you don't have a solid foundation, the house is doomed from the start. People don't realize that these city inspections and the checks of the lending agency are

> in the homeowners' best interest. If they weren't made, some builders would take complete advantage of buyers. After all, most of what a contractor puts into a house is hidden. Only after a person occupies a dwelling for a reasonable period would some of the failings come to light. I make four to five visits to a building site—during the foundation phase, framing, enclosing, and completion.

Don't pay your bill until after the city inspector has given his approval. If the invoice provides a discount within 10 days and net at 30, it makes sense to pay it within the discount period. Otherwise, be sure to make payment within the time allowed.

Lumber

Lumber costs are a major item in the construction of your home—and these costs are increasing every year. A front page article by William Wong emphasized that "Whopping Price Hikes on Lumber Products Wallop Home Builders."

> Just in the past seven weeks, prices on key lumber and plywood products have shot up between 20 percent and 30 percent. Over the past six months, the National Association of Home Builders figures price increases on wood products have added $1,200 to the construction cost of the typical new one-family house. The basic reason for the price surge is sustained demand for wood products from U.S. homebuilders and from foreign customers. The relaxation of controls under the Nixon administration's Phase 3 appears to have encouraged advances, too. And still another factor is a rail-car shortage that has kept some lumber from eastern markets.[1]

With lumber so expensive, it is essential that you as a homebuilder take appropriate precautions to use it wisely and protect it well. Work closely with your carpenter in order to meet his needs. If feasible, order out only the amount required for several days work. Let the lumber company, not your building site, be your storage facility. It is surprising the quantity that can disappear if left unguarded. Other reasons for keeping minimum amounts on your property include the greater risks from damage by vandals; deterioration by prolonged exposure to the elements; and injury to workers and the curious, ever-present youngsters who delight in playing in such surroundings.

[1] Excerpt from the *Wall Street Journal*'s article of March 5, 1973

You should be present to check the amount of lumber actually delivered. Your carpenter should also be with you to confirm that the quality and size are in accordance with what you ordered. If there are any damaged pieces, or other irregularities in the material, be sure that they are returned at once and replacements delivered promptly. Don't sign the delivery slip for any merchandise received at the site until you have checked it carefully.

Each day I made a point to talk with my carpenter to be sure he had all the supplies required. For example, if he doesn't have nails or if they are not the proper size, he may have to stop working until they arrive. There will be such emergencies and by being present you can quickly solve such difficulties and perform mercy missions such as going to the nearby store for ice and soft drinks. There will be errands to run but you must understand exactly what materials are needed. It is disappointing to run five miles for nails and have the carpenter say: "That's not what I wanted; I asked for ten-penny galvanized nails."

Exterior Forms

Once the initial lumber is delivered, the carpenter is prepared to lay out the exterior forms so that the slab can be poured. It also permits the plumbers to come in to do their initial work.

Foundation Plumbing

Now that your exterior forms have been laid out, you can proceed to work with your plumber. He will prepare the foundation plumbing. You may find that the head plumber is the only member of a small firm that has achieved "the master plumber level." He has the most knowledge and skill so it is important that he personally be present during this phase of the project. If trouble should be discovered later, it is terribly expensive to break through concrete that has set for a good period of time.

You will require both galvanized iron pipe and copper tubing. Copper, like lumber, is expensive. Be sure that the copper utilized is the size you specify: don't accept ⅜ inch if you desire ⅝ inch. The amount of pressure and quantity of water for your toilet facilities are dependent upon the diameter of the tubing. Normally, you subcontract with the plumber for the entire job, including the foundation plumbing and all the inside facilities—tubs, commodes, sinks, and hot water heater. Unless you specify your copper size, he will try to keep his costs down by using smaller piping. And even if the specifications call for ⅝ inch, he may substitute a smaller size which could go unnoticed until you wondered why the commodes flushed so poorly.

Check the location of your clean out holes so that if your toilets later become stopped up, they can be readily flushed out. You should also note the location of the outside water valve so that if a leak develops, you can quickly turn off the water supply.

Once the plumber has completed his work, there should be an initial plumbing inspection. Hopefully your plumber has a friendly relationship with the inspector so that it will be done promptly. (I have been informed that to help expedite this procedure, money changes hands at times.)

Grading and Preparation of Foundation and Other Concrete Work

Your next subcontractor will prepare and pour your foundation. This is hard manual labor. Be present when the steel meshing, reinforcing rods, shell material, and other supplies arrive to be sure that you have the quantity needed and that they meet your specifications. The contractor himself normally makes the arrangements to purchase these items and also orders the concrete.

I found it an exciting experience to be present during the actual pouring of the slab. It is important to call the day before to confirm that the concrete will be delivered and that the men arrive on time to lay the slab. This work begins early in the morning and is usually completed that afternoon. Check your weather reports; you cannot pour satisfactorily if it rains. However, a heavy downpour an hour or two after the slab has set will help to cure it.

These workers move at a fast and furious pace. In hot weather, it may be helpful for you to have ice and soft drinks. You may also find it a timesaver to have a "portable john" on the job site. Some workers prefer to take a short break periodically. If gas stations or other facilities are nearby, it may not be necessary to rent one.

Termite Protection and Inspection

Normally you will take out a contract for both the initial termite protection and the annual follow-up service. Be sure that the charges for the annual inspection are no higher than what you would pay a firm on a onetime basis. Your presence on the site at the time the service is rendered will permit you to observe where the openings are located and that each outlet is undamaged so it can be properly capped. The firm's representative can also provide you good advice on signs indicating the presence of termites.

Garage Secured

It is helpful to enclose your garage as soon as possible; this will give you a place to store items and keep them safe overnight. This includes lumber, nails, tools, and other materials. My preference is for a garage completely separated from the house. It means less noise inside the home and, I believe, has a more attractive appearance.

Framing

While the carpenter and his workers are completing the framing, you can meet with other subcontractors and suppliers on the job site. It is advantageous to build during quiet periods of the year or when housing starts are slower than normal. In such situations, people are more anxious to have your business. They will make the extra effort to come visit you and their prices may be more reasonable. We found the lumber salesman made frequent trips to our home and worked with the carpenter in order to preclude shortages of lumber as well as obtain millwork requirements. Millwork refers to the doors, trim, shelving, window sills, and other finishing work. It is helpful to utilize a lumber house that has both the framing lumber and millwork. A personal visit to the lumberyard beforehand will enable you to see the quality of their material. Your carpenter should also have the ability to do the framing and install the millwork.

Exterior Sheathing and Covering for the Roof

You should proceed rapidly to enclose your house to protect it from the elements and provide a measure of security. Sheathing is an initial covering of waterproof material that may be used on the exterior wall of a frame structure. Your description of materials should specify the type required, such as USG gypsum $\frac{1}{2} \times 2 \times 8$ v edge.

Some builders place sheets of plywood at the entrances to keep out intruders and stray animals. Once under cover, you can proceed at a rapid pace by utilizing several subs inside while others are busy outside (see Figure 4-2).

Roof

The roof, like the foundation, should be a long-lasting investment. It pays to use fine materials and employ a roofer who takes great care in his work.

A higher roof permits faster run-off of rain water and less accumulation of snow. Such a roof provides for greater storage space and results in greater warmth in the winter and a cooler interior in the summer. It is also nice to have a high attic where workers can more easily install and maintain your equipment.

Windows and Glass Doors

The less maintenance work you have in the future, the better off you will be. Aluminum windows can be a substantial work saver.

Ample windows will give you a bright and sunny house. A large sliding glass door facing the backyard is also a worthwhile investment. Be sure that the glass door has adequate locks. It is possible to purchase glass that will withstand bricks thrown at it.

Don't install your screens until the painting and other outside work is finished. Otherwise, your screens may be a mess; their replacement may be less costly than trying to clean them. If possible, let your supplier store them for you until needed. If not, your garage may be a satisfactory haven. Cover your windows with paper after installation so that cleaning becomes relatively easy before you move in. Those of you who have used razor blades to scrape paint and plaster from windows can readily appreciate the time and effort required.

Air Conditioning and Heating

Care should be taken to select the appropriate size unit for both heating and cooling your home. In a number of speculation homes, I have found the air conditioning unit to be undersized and ineffective on very warm days.

We installed two 3-ton units; the second unit kicks on automatically when required. We also put in an electronic filter system. It does a good job of removing dust, smoke, and other particles in the air.

It pays to buy a fine quality heating and cooling unit from a reliable dealer. Check to see that appropriate insulation is used to wrap the air ducts and that they are of adequate size. The outside air conditioning unit should be located so as not to present an unsightly appearance or be such a distance from the house as to reduce its efficiency.

Electric Turbine

An electric turbine will go on and off automatically at present temperatures. The purpose of the turbine is to reduce the heat in your attic. It pays to buy a good one that will give reliable service over a long period of time. More reasonably priced turbines are wind driven but they are not as efficient. Normally a turbine should be placed to the rear of the roof so as not to detract from the appearance of your home.

Electrical Wiring

If feasible, have your city's electric utility company visit you to obtain their estimate as to the appropriate lighting needs for your home, including the number of outlets. Naturally, they are interested in maximizing electrical usage because it will give them more business. Nevertheless, it gives you a good basis for talking with your electrician. With the continuing increase in new electrical products, it's safer to have more rather than less outlets.

Be present while the wiring is being installed. If you ask for number twelve wire for both leads and returns, it is wise for you to personally see that this is accomplished. Although your electrical inspector should pick up such points as wiring size and sufficiency of outlets, this isn't always the way.

Telephone

The telephone company will be happy to work with you in determining the type of equipment best suited to meet your requirements. Be liberal in your estimate of future needs. For example, I put outlets in the living room, dining room, den, and kitchen areas as well as all the bedrooms. It proved to be a good investment with teenagers in the house.

Insulation

The best time to make certain that you have adequate insulation is during the building phase. Be sure that the material received meets the specifications contracted for with your supplier and that the workers place it in the areas difficult to reach. I found that by staying with them they filled in a number of areas that would have otherwise been overlooked. The following comments by the local utility company may be helpful to you:

> In homes with little or no insulation, much of the heating generated by the heating system is quickly lost, thereby increasing heating costs. Adequate insulation, however, properly installed (especially in ceilings, as warm air rises) helps keep the heat in and the cold out, to let you enjoy maximum comfort, operating efficiency, and economy from your system. Insulation should be the equivalent of 6 inches of rockwool or fiberglass in ceilings, 4 inches in exposed walls, and 2 inches in floors of homes without slab foundations. Also, be sure to seal cracks around

Figure 4-3 The Stillman house built in two months at a 25 percent savings.

doors and windows with weatherstripping or caulking compound to prevent heated air from escaping.[2]

The Stillman House

As an example, let's examine photographs of our house that will help to elaborate other items listed on the PERT chart. We will point out features in each photo that may be of interest to you in building your own home. You might also wish to refer to the blueprints of our house that appear in Appendix I.

Please keep in mind that these comments reflect our views and desires. You must build and furnish a home to suit your own taste and temperament. This rare occasion to express yourself is one reason why I don't believe prefabricated homes can ever be adequate substitutes for custom-built dwellings. In America, we have seen craftsmanship give way to mass production in almost every field of endeavor. The last great exception is in the homebuilding industry. This privilege should not be sacrificed on the altar of cost and conform-

[2] "Insulation More Important Now Than Ever," *Homemaking*, New Orleans Public Service, Middle South Utility System, March 1973. (Courtesy of New Orleans Public Service, Inc.)

ity. Furthermore, by subcontracting and utilizing sound management concepts, costs can be held to within the competitive range of prefabricated homes of a comparable size and design.

Brick

We selected a New Orleans hard tan brick that would give the home a distinctive appearance. Old bricks are available in most communities, since wrecking crews are frequently tearing down buildings and enterprising businessmen purchase them for resale after appropriate cleaning and sorting. Some people buy cheaper bricks and then go to the expense of painting them. In my view, this is not as attractive and maintenance is costly.

We extended our brick across the entire width of the house in order to give a continuous flow in preparing the driveway and side gates. This interweaving effect also provides added strength.

Driveway and Entrance Way

In order to add a bit of distinctiveness, we used washed gravel. The drive is wide enough to accommodate two vehicles and the entrance wide enough for two people to walk abreast. We also found it helpful to have the walkway extend to the street.

Garbage Receptacles

Two receptacles were necessary, considering the volume of trash that accumulates. It helps to surround the immediate area with concrete rather than to place the receptacles on a grassy knoll. Be sure that you have a porous substance at the bottom so that there is adequate drainage when it rains.

Entryway

We wanted a slate floor at the entrance. It is easy to maintain and is attractive. The double doors make it easy to move large items in and out.

Yard

Simplicity was our theme with respect to the lawn and shrubs. We checked with the state and city agriculture departments to find the most suitable plants. The oak tree is a present from the people who developed this area. Landscaping your home will be discussed in Chapter 9.

Figure 4-4 Driveway and garage.

Driveway

We put a lamp post at each end of our property to provide lighting and add a distinctive flavor. The drive itself is wide enough to permit a large car to move in and out with ease. With two teenagers driving the family car, it has paid to have this ample width.

Fence

We decided to use wood rather than a brick fence and subcontracted for first quality cedar wood. We felt that it would provide a more spacious and country atmosphere than if we used a brick similar to our house. If you use wood, check carefully to see that there are a minimum of knotholes. The fence itself should be solid enough to withstand very strong winds. This can be accomplished by placing the posts close enough together and securing each post in a solid foundation. It also helps to use three, rather than two, lateral boards for support. Keep in mind that labor costs are virtually the same for a shoddy fence as for the finest, but inferior materials will result in a shorter life and labor costs will then be required again. Stick with good quality in fencing and all your other needs.

Eaves

We put our eaves in the ground so there would be rapid drain-off of water. We also found a wide overhang (24 inches) provides more shade and helps shield us in inclement weather.

Garage

As you look at the photo, you will see the garage door is in line with the driveway. Time and again we have observed homes that do not have a direct entry to the garage because they did not locate it on the property line. Several of our neighbors failed to do this; as a result, they cannot safely get a second car in and out of their garage. Those that try find they have frequent damage to car and home. With land so expensive today, it pays to use every bit of your property to its best advantage.

What type of shelter should you have for your car? Many homes built for speculation have carports. An enclosed garage costs more but has definite advantages. It reduces vandalism and permits you to safely store items such as your bicycle, lawn mower, and tools. It better protects the car from the elements and can help eliminate starting problems.

We purchased a solid, extra wide garage door so that there would be no difficulty getting in and out with two cars. I am lazy by nature and find one of the comforts of life is an automatic door opener.

Figure 4-5 Garage entryway.

Garage Entryway

Figure 4-5 gives you an idea of the spacious area of our garage. It is ideal for table tennis; there is even enough room to play in inclement weather with the door closed.

Living Room

We wanted a spacious living room that could be used for entertaining and daily living. Our idea was to build a home where we would use all the rooms. Living in Paris for three years, we found that many of our French friends only used their living rooms on very special occasions. In New Orleans, a number of homes have very small living rooms and families spend most of the time in their dens. In our case, with two teenagers, we found it best to have two entertaining areas so that each of us could be with our own friends. Doors close off the living area and adequate insulation keeps the noise level to a minimum.

Figure 4-6 Living room.

Figure 4-7 Dining room.

Dining Room

We chose a separate dining room with sufficient space to accommodate twelve people for dinner. Louvered doors can be closed to separate the living and dining areas. We find this partition especially nice for surprise parties, birthdays, and other special events; then we can use the living area for cocktails prior to dinner.

Note the built-in china closet that can be seen in the rear of the photo. We find this is a convenient space-saver. It is also a cost-saver that results in one less piece of expensive furniture. A dimmer switch adds a nice touch and can be installed during the building phase at little extra cost.

Kitchen and Breakfast Nook

Our objective was to have a bright and airy kitchen with an adjoining breakfast nook. We put both fluorescent and regular lights in the kitchen. We placed our breakfast nook where we could receive the sunlight in the morning and built a little alcove as an added flair. The light in this area can be lowered and raised as desired.

Figure 4-8 Kitchen and breakfast nook.

My wife stands guard over the kitchen domain. As you will note, it is a compact area. We wanted to save steps and place the shelving within easy reaching distance. Behind my wife is a storage area enclosed by a louvered door.

We selected an electric kitchen and found the self-cleaning oven a time-saving item. Another kitchen appliance we enjoy is the self-defrosting refrigerator with ice-maker.

Figure 4-9 shows another view of the kitchen indicating the location of the window, sink, shelves, refrigerator, and counter space.

The floors of our kitchen, breakfast nook, utility room, den, and hallways are of terrazzo, "a mosaic flooring made by embedding small pieces of marble or granite in marble and polishing," according to Webster. We found its appearance most attractive and the upkeep minimal. In checking some other homes, we found numerous cracks in the terrazzo. We have eliminated that problem by making the floor thick enough (2¼ inches) and having it unbonded, that is, separated from the concrete slab. This type flooring takes skilled craftsmen and requires considerable time. The workers came on three

Figure 4-9 Kitchen—a closer look.

separate occasions to complete the job. There are other good floorings which are less expensive and easier to install. You may like a softer composition that will be kinder to the feet and less destructive to breakable objects accidentally dropped.

Figure 4-10 View of den looking into breakfast nook.

Den

The den receives the most use because here is where both our library and television set are located. I hate to admit which one receives the greater usage. Again, our wish was to have a bright and airy room, so we utilized light-colored paneling and a sliding glass door with three large windowed sections.

Our books are meaningful to us and so we built ample shelving for this purpose. It also provided us a chance to display bric-a-brac that we acquired in our travels throughout the United States, Europe, and the Far East. As you can see in the back of the photo, we have double doors that can close off the den if so desired. Beyond the double doors to the left, we built a center hall closet to take care of guests' wraps. You may also note a sliding panel door. We placed several throughout the house and found these pocket doors to be real space-savers. A word of caution—be sure they are installed properly, because they are expensive to repair.

Patio

The wood fence to the rear of our lot and the brick garages of our neighbors provide privacy in our patio area and a pleasant view from our den. I cannot stress too strongly the importance of checking out the area carefully prior to buying your lot. We like the idea of skinny-dipping, and our location permits this pleasure. Beyond the sliding glass door in the den, we built an 8-foot overhang so that we could entertain outside and still have some protection from the sun and rain.

Master Bedroom

The bedroom wing is located on one side of the house as can be seen from the architectural drawings in the appendix. We have four bedrooms in this area, including two complete baths with shower and tub. This area can be closed off from the remainder of the house by two sliding pocket doors. We

Figure 4-11 View of den from the patio.

Figure 4-12 Master bedroom.

put the master bedroom to the rear of the house for privacy and quiet. In the back of Figure 4-12, you can see the bath and commode. Comparable space to the right rear (see Appendix I) is used for a walk-in closet. Between the commode and closet is a large wash basin with drawers to accommodate necessary bath supplies. Sliding pocket doors seal off this area and it makes a nice dressing room. Planning your home, you must allow for ample closet space.

Study

Two of the other bedrooms are used by our teenagers. I converted the fourth one at the front of the house to a full-time study. As Figure 4-13 indicates, the closet is of sufficient size to hold a large filing cabinet; built-in shelves give me adequate space for my reference books. If we were to sell our house, there would be no problem in changing this to another bedroom. We considered having an extra bedroom exclusively for guests. But space is too valuable today to set aside special areas for this purpose. We can convert the study and/or living room into sleeping quarters for our friends or relatives.

Main Bathroom

The main bathroom is situated between the two teenagers' bedrooms. We liked the idea of separating the commode from the wash basins. It was also helpful to install two wash basins because of the heavy traffic in the bathroom each morning.

In our bathtub, we used a built-in non-slip material rather than a rubber mat. We also found it more convenient to have a shower curtain than built-in shower doors. While looking at homes, I thought the curtain was the cheaper way and that was why so many used it. Although it is less costly, it has a more important advantage–it is a great deal easier to keep clean.

We like ample lighting in our bathroom. In addition to a window, we have fluorescent lighting and three other fixtures. This includes a heat light that not only brightens the room but keeps it warm in the winter.

White was selected for our fixtures because it can be used with almost any combination of rugs, towels, and accessories; and it gives a cheerful appear-

Figure 4-13 Study room.

Figure 4-14 Main bathroom.

ance. It is also slightly cheaper to purchase white fixtures. In view of the fact that you may sell your home one day, it helps to have colors appeal to a majority of potential buyers.

Utility Room

The utility room is where the washer, dryer, and hot water heater are located. We also have a large sink which can be used for messy jobs, such as cleaning fish. An enclosed half-bath has proved to be a real convenience in this area. The utility room also has a door to seal it off from the breakfast nook.

Attic

The attic can be a splendid storage facility and ample flooring should be installed. Items accumulate rapidly; and if you expect to use your attic frequently, it should be of adequate height, room, and light.

Figure 4-15 Utility room.

Figure 4-16 Attic.

Cleaning Up

Many building sites are real eyesores and a nightmare for environmentalists. Trash is strewn everywhere with little regard to the neighbors' property or people who use the adjacent sidewalk. Materials may even be piled in the street.

You, as builder, can do something from the start to indicate your pride in ownership. For example, a retired naval officer, who did his own subcontracting, purchased several trees after buying his lot. He also placed large receptacles at strategic locations on his property. Each subcontractor was asked to have his men place their debris in these containers. At intervals, he had a truck haul away all the rubbish. It cost him $5 to $10 a load, depending upon the quantity. He made a point to sweep up the premises daily and neatly arrange the materials. This helped him to keep an accurate inventory and permitted him to instruct his suppliers to furnish additional items before he ran out. The result: workers made an effort to keep the premises clean and they liked working in such surroundings. It also helped keep the injury rate at zero and losses from vandalism to a minimum. Move-in time was speeded up because it meant no massive clean-up job.

One other advantage—he had no insect problem on first moving into his home. A surpising amount of debris may be enclosed in walls, attics, and basements. It's little wonder that nests develop and odd creatures make nocturnal visits.

Summary

The building phase is the action phase. During this period, your prime responsibility is to be on the job site. The results will depend in large measure upon the amount of personal supervision that you provide. You may have the best plans in the world but it is necessary for you to see that they are implemented properly. If you cannot be present, then your wife or someone you can depend on should take your place. It is this personal attention, using sound management principles, that gives the subcontracted house a competitive edge—in quality and cost. No contractor can afford to provide this caliber of supervision.

During the building phase, you may wish to use a PERT chart and timetable. It helps to have a good working relationship with your subcontractors and suppliers.

Time is valuable so employ it wisely. This may include running errands for your subs to preclude a work stoppage and having suppliers visit you on the site.

Keep your premises neat; it helps to reduce injury, increases pride in workmanship, and assists in maintaining better inventory control.

CHAPTER 5
Checking on What You Have Done

While walking to my office recently, I observed three homes under construction. In each case, the work had come to a standstill. In the first instance the mortar was not available for the bricklayer, so he took another job. At the second house, the electrical inspection had not been made. The contractor had failed to contact the responsible city department. At the third site, there was a conflict between the carpenter and the builder over the terms of payment.

These delays could have been prevented if the builders had taken adequate control measures. Prior to examining the specific checks which can be made, let's put the control function in proper perspective. It should be apparent from reading the previous chapters that the three functions of planning, organizing, and controlling proceed simultaneously and are a continuous process. The process itself follows a repetitive and on-going pattern of *planning* what you are going to do, *doing* what you planned, and *checking* to see that it is done according to the plan. For example, let us assume you *plan* to clean up the premises at 7:30 A.M. While tidying up (*doing*) you uncover damaged lumber (*checking*). You should then devise a *plan* for its return and replacement. Concurrently, other subcontractors working for you would be performing their various planning, organizing and controlling functions. Thus, it is evident that you as manager cannot perform each of these functions in isolation. You might conceive of them as part of your management bag of tools that must be utilized harmoniously to get the job done efficiently.

Specific Controls and Inspections

In order to determine the degree of effectiveness with respect to the two major functions of planning and organizing, you must apply certain control techniques. This quantitative and qualitative data permits an insight into the standard of performance and enables the homebuilder to take remedial measures where appropriate. Let us look at specific controls you as a homebuilder might use.

On-the-Spot Inspections

Being at the job site, you have an opportunity to immediately check both the quality of work and the materials being utilized. For example, you may be particular about the color of paint desired in each room. Your presence will permit you to inform your painter after the first few strokes of the brush that it isn't the proper pigment. It is much better to discover such a mistake quickly than to find out after the entire interior of the house has been painted.

You can also check, upon delivery, such supplies as material for your fence. If your inspection reveals a number of boards with large knot holes and inferior quality, you can demand replacements. In contrast, if the fence is installed before the deficiency is discovered, it is more difficult to convince the subcontractor to replace it.

Being on the job enables you to prevent any unnecessary delays. If a subcontractor is absent, you can call his home or office to find out the reason and can arrange for other work to be accomplished. Likewise, if supplies are not delivered on schedule, you can take action to obtain them—either by placing a phone call requesting prompt delivery or picking them up yourself. You will also have a chance to talk with your subcontractors to recheck schedules and preclude delays.

The significance homebuyers are placing on inspections was discussed in a report by Jeffrey A. Tannenbaum entitled "Homeowners Outraged by New House Defects and Delays in Repairs."

> Prospective buyers also are becoming more cautious about choosing a home in the first place. Many, for example, are having new houses inspected by Mr. Heyn's inspection service before closing contracts. Mr. Heyn, who founded the service in 1969, says he had expected most of his customers to be people buying 20 to 30-year-old houses. "The biggest surprise is that over 50 percent of our business is new-house inspection," he says.
>
> Most of the complaints around the country involve houses constructed by production builders, those who build from a number of basic designs. But even buyers who use custom builders are becoming more skeptical. Roland Eppley, president of Eastern States Bankcard Corp., had Mr. Heyn's men inspect his custom house in Manhasset, N.Y., during four different stages of construction.

> Mr. Eppley, who paid more than $100,000 for the house, says the added $260 he paid Mr. Heyn's firm "gave me peace of mind."[1]

Imagine how effective your control could be with continual on-the-job inspections in contrast to those made on four separate occasions.

Other Inspections

Visits from loan agency officials as well as electrical, plumbing, and other city inspectors can provide splendid means of checking the quality of workmanship and material. I found they took a greater interest in making thorough inspections when I was present. Even when they approve the work, it can be helpful to ask their advice on what could have been done to make it better.

If irregularities should be found, take prompt action to correct them. This should be accomplished in a friendly working relationship with your subcontractors.

Another inspection I found helpful was visits from my friends. I asked their views on workmanship and materials.

An important inspection is made when the house is available for occupancy. You should plan to move in within 72 hours after completion. If necessary, some work can be finished while you are in the house. The advantage of being in the home is that you have a 24-hour check on all aspects of work. Be sure to check *everything* with care. If it is winter time, do not hesitate to try the cooling system.

In lieu of the after move-in inspection the following ditty of Colleen Bare points up the ideal solution to obtaining a house without flaws.

House Plan

If you are interested in the cause
Of building a house without any flaws,
There is only one way that's definite:
Before you build it, you must live in it.[2]

Meetings

Each morning, I met with the subcontractors upon their arrival. We discussed what was to be accomplished and any problems which needed to be

[1] Excerpts from the *Wall Street Journal*'s article of April 3, 1973 © 1973 Dow Jones & Company, Inc. All rights reserved.

[2] Pepper and Salt, the *Wall Street Journal*, March 28, 1973, p. 12.

resolved. Prior to their departure, we confirmed the time they would be at work the next day.

As a result of these sessions, the subcontractors would tell their workers of any changes. It also meant that we were all informed of matters of common interest and problems which might arise.

Meetings also offered the opportunity to exchange ideas. There is much to be learned from these skilled craftsmen *if you are willing to listen.* They have many years of experience and you can acquire considerable knowledge about various aspects of homebuilding. I have found they are proud of their own labor and critical of other workmanship. By having them point out deficiencies, it enables you to be better armed with facts when requesting corrective action.

In order to properly prepare for these daily sessions, it is essential for you to do your homework the night before. Review your PERT chart, plans, description of materials, timetable, and budget. Their time is valuable and normally subcontractors work right along with their people. These sessions should be brief, and ideally you should conduct your business while they are working.

Topics that come up frequently include interpretation of the architect's drawing, deadlines to be met, materials needed, inspection requirements, and payments. Try to provide an atmosphere that will encourage the subs to express their views.

Comments of workers were also appreciated. I found some of the men would arrive before their boss and this provided an opportunity to learn about home construction.

Another good time to rap with the subs and their workers is during lunch breaks. They're relaxed and are usually a congenial group. You should provide a climate that encourages them to present their views on what can be done to make it a better house. These lunch gatherings are a good time to brainstorm a new idea or propose a modification. By brainstorming, I mean obtaining the maximum number of ideas without regard to the quality of the suggestion. You might next proceed to obtain the most feasible proposals and finally select what you consider the best recommendation. For example, we discussed changing the pitch of the roof, increasing the size of the breakfast nook, and using a new plastic material. These sessions may help you arrive at sound decisions.

Meeting with workers also provides other benefits. Once I was informed of an article on home construction that appeared in their trade journal. On another occasion, I was told of a house being constructed in the city that was using a new flooring material.

Payment by Check

It is wise to make the vast majority of payments by check and to obtain necessary receipts. Obviously, you should pay cash only when buying a few soft drinks or several screws; but do keep a record because these small amounts add up.

It is a good idea to endorse the back of the check, indicating the purpose of the payment: "Final payment for painting home at 2311 Oriole, per contract #201."

Prior to return of your monthly statement, you may need to show your loan agency that payment has been made. Your receipt will serve this purpose. It can be valid even if written on a scrap of paper as long as it is signed by a responsible person in the organization. Take advantage of any discount offered for early payment; otherwise, wait until the due date.

Don't hesitate to stop payment on a check if you find that material is defective or work is not accomplished as promised. It may be the best way to see that appropriate corrective action is taken. Money is the best incentive for getting a job done. Once payments are completed, you have lost your bargaining power. I recommend 10 percent be withheld on major jobs until you make a final inspection.

Budget

Your budget can serve as a planning and control document (see the budget example in Figure 2-8). The initial columns of figures indicate the several estimates of what the various items cost. Their total provides you with an indication of the money required to build your home. This information enables you to talk intelligently with your lending agency.

Your estimate, however, does not always coincide with actual costs. In most cases, cost exceeds the amount planned for in the budget. This may be true even though you have price agreements with your subcontractors. Supplies are a different story. The sharp rise in lumber costs, for example, was pointed out in Chapter 4. There are also changes in plans that can be costly. However, if you use your budget effectively as a control device, this variation between estimate and actual can be minimized by promptly posting the actual price paid for each item. You should then check carefully to see if you have exceeded your estimate and by what amount. The real value of budget control will occur only if you take positive action to correct the problem.

We shall assume that after the foundation work has been completed, you find that your budget estimate has been exceeded by 5 percent. The problem is obvious—you spent too much. You must then decide what to do about it.

After obtaining the facts, you proceed in management terminology to look at alternatives. One solution would be to ask your loan agency for more money on the basis that the cost of the entire project will increase by 5 percent. A second alternative might be to substitute lesser quality materials on the remaining construction in order to make up the difference. A third possibility could be to eliminate something from the house. From the various alternatives, you can then proceed to make a sound decision. This approach follows the sequence presented in Chapter 1: determine the problem, obtain the background facts, look at the alternatives, decide which alternative to implement.

Filing System

A good filing system is a timesaver and can be an effective control technique. If organized properly, it permits you to find records quickly. Such information is essential at times in order to make wise decisions. I kept my records in manila folders in a file cabinet. Each subcontractor and supplier had an individual file. Likewise, checks, budgets, PERT charts, photographs, and so on were filed separately in alphabetical order for easy reference. Keep your filing system simple.

PERT

PERT was discussed in CHAPTERS 2 and 4. It is a splendid planning and control document and should be followed with care during the building phase.

Your budget serves as a money control whereas PERT permits you to keep your construction on schedule. PERT initially enables you to determine how long it will take to complete the job. It also points out the numerous tasks that can be accomplished concurrently in order to minimize completion time. For instance, if a strike should occur in the cement industry and your schedule shows it is time to lay the driveway, your PERT chart can assist you in selecting a feasible alternative. Regardless of how well conceived, a planning document like PERT, cannot foresee all eventualities.

PERT is only as valuable as you make it. If, for example, you set it up on an unrealistic basis, it is foolish to waste your time on such a project.

Photographs

My Minox® camera was a useful control device. It was small enough to carry in my pocket and it enabled me to quickly make a permanent record of any potential problem. For instance, if there was a question about faulty lumber at the time of delivery, I took a picture of it with the company truck in the background, or I might include in the photo the man who made the delivery.

If you intend to do any future building, photos can be helpful in going over the step-by-step process. An analysis will enable you to make improvements the next time around. I would like to have taken movies of the entire home construction process. After all, sports teams make good use of film to analyze strengths and weaknesses of their own team and their opponents. Management's objective is to make a healthy profit; to do this, it is usually necessary to have a winning team.

Communication

It is so important to talk with your workers, suppliers, subcontractors, and other homebuilders. The information you receive can be valuable.

When you give instructions, be sure that what you say is clearly understood. One way to find out is to have the person repeat back what you told him. Keep in mind that some workers and suppliers may not have your education and background. They use different terms and may interpret things differently. For example, my plumber kept talking about the "zinc." I couldn't understand how he was going to use such a protective coating in my house. That evening, I rechecked the dictionary. The next day he pointed to our kitchen "zinc" and I saw that it was our "sink." If you have any questions regarding interpretation or misunderstanding, put it in writing.

Legal Action

Another control measure is positive follow-up action, but take only the necessary amount of action required to get the job done. If a subcontractor or supplier has failed to meet his commitment, you may wish to proceed as follows:

1. Call the person and courteously explain that the work was not satisfactory and set a time for it to be corrected. By having the contract in writing, you have evidence to remind him of the agreement terms.

2. Make a second phone call firmly explaining that you expect the work to be done and set a second deadline.

3. Write a letter stating you will take necessary legal action if the work is not done within a specified time. Send copies of letters to appropriate parties such as a better business bureau, a local consumer affairs office, Ralph Nader, and the White House Office of Consumer Affairs.

4. Proceed with the necessary legal action. You may find the small claims court is sufficient to obtain the amount of damages due you. If not, secure the services of a highly respected lawyer. Your chances of being successful in such a situation are improving. The "buyer beware" doctrine is no longer as valid. This change is pointed out in the following article in which a homeowner threatens to sue his builder:

> Mr.——— says he'll sue if necessary. In the past, most courts have routinely sided with builders, citing the "buyer beware" doctrine. But the Research Institute of America, a business advisory service in New York, says that courts in 19 states have changed, often ruling in the homeowner's favor in suits over such things as faulty wiring, poor grading, or badly installed ductwork.[3]

It is wise to take action only for matters of some value. Don't waste your time and energy quibbling over minor points to satisfy your ego. Instead, get on with the job. But once you decide something is important, proceed with vigor to gain a fair settlement. In theory, your careful planning and checking on both workers and suppliers should preclude having to take such drastic steps. However, it doesn't always work that way. Occasionally, someone in whom you have great trust and confidence, may let you down.

Summary

The final management tool I used was control; that is, I checked on what was done to see if the workmanship and materials met the standards agreed to in the individual contracts. By being on the job, you have an opportunity to examine supplies as they arrive and work as it is completed. It is important to have a written agreement with each subcontractor. For example, if number 12 copper wire is required for leads and returns (in accordance with the electrical quotation), you don't want to find number 14 being used. In addition to

[3] Excerpts from the *Wall Street Journal*'s article of April 3, 1973 © 1973 Dow Jones & Company, Inc. All rights reserved.

quality control through personal inspection, you must also check the budget and PERT.

Bills submitted should be in accord with expenditures agreed upon. You should withhold final payments (about 10 percent) until the material or project can be tested or inspected. A decided advantage to moving in promptly upon completion is that you get a 24-hour check on workmanship and equipment. It is much easier to secure service before complete payment is made.

PERT served me as a splendid means of time control. I had several unforeseen emergencies and was able to reallocate resources thanks to this concept. The major bottleneck was a strike by cement truckers while the bricking, driveway, and pool were in progress. PERT permitted me to revise my schedule of subs while alternatives were found to provide mortar and cement. This reallocation kept the project on time.

Never pay in advance. I made a mistake by paying half the pool price the first day that the excavation equipment and crew arrived. The men demanded that I pay them before they began work; their past experiences showed them they wouldn't be paid promptly. Fortunately, I was able to stop payment on my check and I learned a good lesson cheap.

CHAPTER 6
Do You Want a Swimming Pool?

There are nearly one million swimming pools currently on the property of homeowners throughout the United States. In some wealthy residential areas, you will find a permanent pool in virtually every backyard. The initial cost approximates that of a luxury automobile.

You should give considerable thought to determining if you really want a swimming pool. This decision should be arrived at during the planning phase of homebuilding. Here are some of the pros and cons that may help you reach your decision.

Advantages

Exercise

Swimming is a delightful and invigorating exercise for every member of the family. A pool allows you to develop athletic skills such as lifesaving, scuba diving, and water polo.

Entertainment

It can provide an enjoyable way to entertain relatives and friends. You can barbecue nearby and spend an afternoon or evening of mild exercise and good conversation in a picnic environment.

Increased Property Value

A pool could increase the value of your home more than its original cost.

Status symbol

Here is one way to keep up with the Joneses if you feel the need.

Aesthetics

A pool can be blended into a tastefully landscaped yard that permits a pleasant view from your patio and other rooms. There is also the satisfaction of creating your own design.

Cost Savings

By building a pool at the time you construct your home, you may realize a substantial savings in the initial price as well as interest charges.

Other savings

If you and your family use a swim club or other facilities, there could be dollar savings here. Our neighbor dropped his club membership when he built his pool and indicated it saved him considerable travel time.

Pride in Ownership

There is satisfaction in knowing that the owner can do as he pleases about its use, maintenance, and decor.

Disadvantages

Cost

It is a heavy initial expense of $5,000 to $10,000, at a time when you are burdened with many other construction costs. It is possible to buy an inexpensive pool that can be placed above the ground but I am speaking about a quality in-the-ground pool.

In addition to the cost of the pool itself, you must consider the decking, backyard lighting, landscaping, and fencing.

After the pool is built, there will be maintenance expenses which may vary from $200 to $1,000 a year, depending upon how much work you do yourself, the size of the pool, the kind of repairs, where you live, and climatic conditions. Let us summarize our costs:

Acid and algaecide	$ 14
Chlorine	80
Diatomaceous earth	8
Electricity	120
Pool equipment	50
Repairs	100
Total Annual Cost	$372

A weekly cleaning service would cost $450 a year and a pool heater adds approximately $80 to the gas bill. Thus, our yearly expenses could have been increased to $902.

Upkeep

Once your pool is built, you have taken on an additional responsibility. If you wish to have someone else maintain it, you must find a reliable firm and see that the work is done properly. Even with a maintenance contract there is day-to-day upkeep, insurance, and safety precautions which must be taken.

Accidents

A pool can be the cause of numerous, and even fatal, injuries. We are always concerned when youngsters swim in our pool. Be sure to have appropriate insurance coverage, first aid equipment, and a knowledge of artificial respiration.

Useability

Will you really use and enjoy your pool? If you find that it is a worry, forget it. The novelty wears off quickly and the composition of your family will change over the years. Also, you may prefer the congeniality of a club or public swimming facility.

Financial Loss

Your pool may be a drawback when you go to sell your home. A number of people don't like them or feel they need one. A factor in making your decision is whether or not they are prominent in your neighborhood.

A Managerial Approach

If you decide to build a swimming pool, follow the same management concepts presented in the previous chapters on the construction of a home. Your objective might be "to have a quality swimming pool built within 30 days at a minimum cost." You must plan it wisely, have it built in accordance with your plan, and make adequate checks to ensure construction as specified in the contract. Achieving your objective requires the efficient use of three resources—men, money, and materials. You should be in a good position to make sound decisions by delineating your objective, determining the resources available, and understanding your managerial functions.

Figure 6-1 Neighbor's brick garage provides privacy and pleasant view from pool.

Selecting Your Lot

The ideal time to make the decision to have a swimming pool is prior to buying your land. It is advantageous to have a pool situated so that it affords privacy. We considered it important to select a lot where our neighbors could not look directly into our backyard and where buildings facing our patio area were attractive. Figure 6-1 points out the brick garages of our neighbors that provide both privacy and a pleasant view.

In selecting your lot, also take into consideration its shape and size. Will its dimensions be adequate for the type of pool you desire?

Financing Your Pool

If you build your swimming pool at the time you construct your home, it becomes part of the total financial package. Thus, if the current mortgage rate is 7½ percent, this is what it will cost to finance your pool. If you wait, however, until after the house is completed, you must obtain a home improvement loan. The difference is approximately 2 percent, based upon charges of a local savings and loan association. In contrast to the savings and loan 9½ percent charge, a finance company advertised real estate loans at 14.1 percent.

The ad read: "Homeowners, now you can get $3,000 to $30,000 when secured by a first or second mortgage."[1] Their rates were spelled out as follows:

Amount financed	Monthly payment	Months to pay	Total of payments	Annual percentage rate
$3,000	$ 70.00	60	$4,200.00	14.1%
$5,000	$116.66	60	$6,999.60	14.1%
$7,000	$163.33	60	$9,799.80	14.1%

If you select a finance company, your loan must normally be secured by a mortgage to obtain their lower rate. Otherwise, it could cost up to 42 percent. Length of payment is listed as five years in contrast to the much longer loan period if you build the pool at the time the house is constructed. Note also how interest charges mount up; if you borrow $7,000 for 60 months, it costs you $2,799.80.[2] My recommendation is that if you decide to wait, save your money and pay cash for the pool.

Other Considerations

Once you have determined your financial needs and found a desirable lot, there are other factors to consider.

Basic Planning

Plan your pool concurrently with the rest of your home and include it in the initial house sketch. To help you develop a suitable drawing, look at a number of other home pools. Current articles and books on the subject can inform you of latest developments. Be sure to consider your pool as part of the total home picture. That is, relate it to your landscaping, patio, fence, garage, and the house itself.

[1] "Associates Financial Services of America, Inc.," *Times Picayune*, May 22, 1973, p. 5.

[2] You may be asked to take out a life insurance policy for $534 (if not, you would be required to assign a $7,000 portion of your present policy for the life of the loan). An accident and health policy is encouraged amounting to $602. Thus, your cost could be increased to $3,936.

Selection of a Builder

Spend considerable time and study prior to selecting a company to build your pool.[3] Contact experienced people; and obtain their views on the quality of construction, costs, time required, and the reliability of the firm. It pays to call your bank, better business bureau, and the local consumer affairs agency. You should also find out if they belong to the National Swimming Pool Institute.[4]

Make a point to visit the offices of pool builders. You can get a good idea of the operation by observing their place of work. Also visit homes where pools are under construction and talk with the workers. Note the quality of their workmanship. It requires good supervision and skilled craftsmen to build a quality pool.

You also should ask pool builders if they will furnish you with a certificate of insurance from a recognized company. The certificate will indicate the kind of insurance, expiration date, and limits of liability. Insurance coverage may include workmen's compensation, employer's liability, bodily injury (except car), property damage, and automobile liability. In the certificate will be a statement to the effect that "in the event of cancellation of said policies, the company will make reasonable effort to send notice of such cancellation to the holder at the address shown herein, but the company assumes no responsibility for any mistakes or for failure to give such notice." Another means of protection is to have the company obtain a bond. Thus, if the work is not in accord with the contract, you would have an insurance agreement covering a stated financial loss.

Where possible, make your selection from three competent firms. Determine which one will give you the best deal, including quality, price, and service. It will be necessary for you to furnish them with specifications and a preliminary sketch to ensure they are bidding on the type of pool you want. Figure 6-2 is the drawing, based upon our initial design, developed by the company we selected.

[3] You may wish to consider subcontracting your pool in a manner comparable to your home. If so, the procedures outlined in Chapters 1 to 5 should be helpful. You would utilize PERT, timetable, and the other concepts presented. My recommendation is that you do not subcontract because your time can best be employed in the supervision of your entire home, including the pool.

[4] This organization can be helpful in providing you information and literature on pools. Their address is: 2000 K Street, N.W., Washington, D.C. 20006.

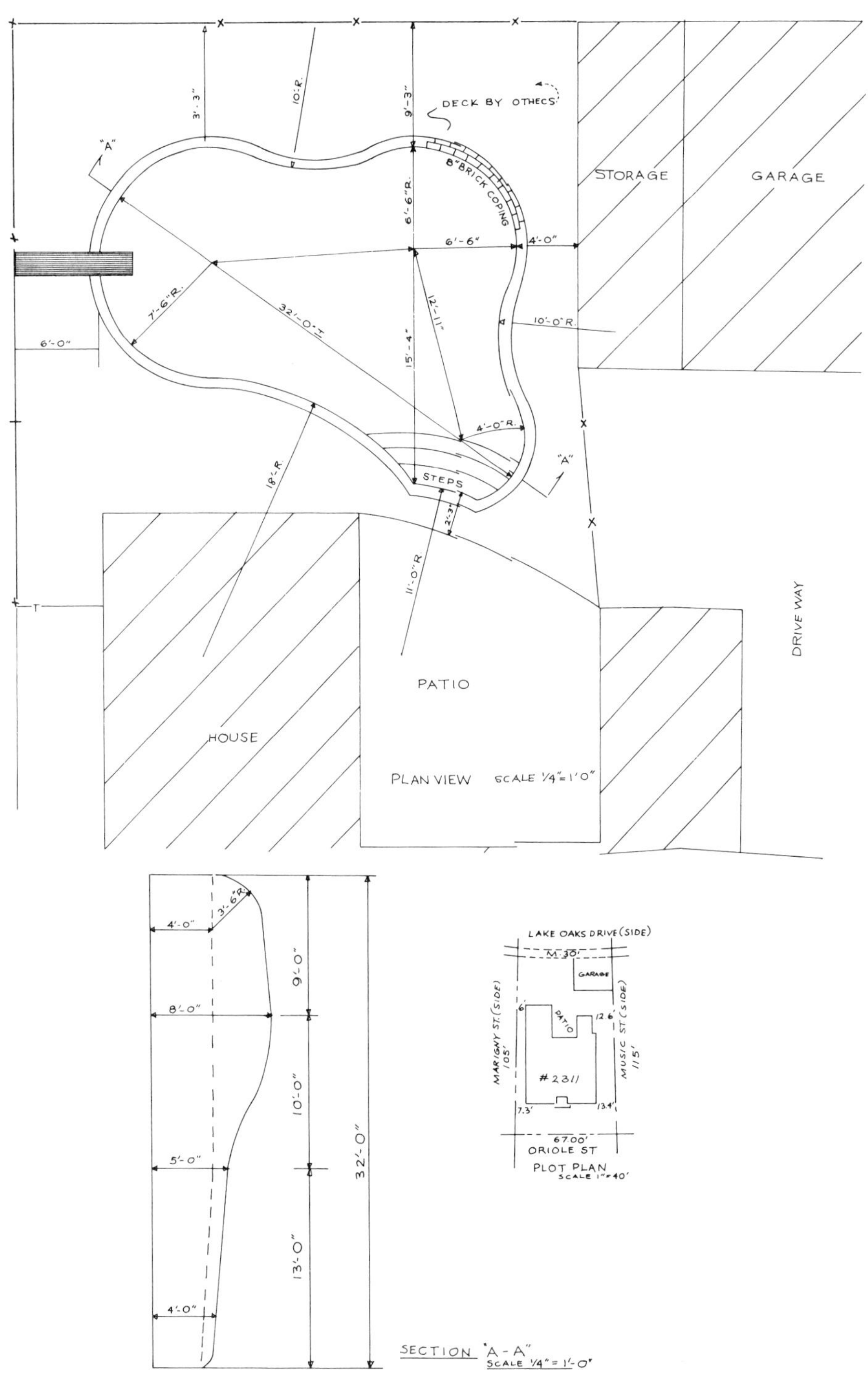

Figure 6-2 Blueprint drawing of swimming pool based on Stillman's specifications and initial design.

Specifications

A good sketch and detailed specifications should enable you to receive accurate bids which will be helpful in selecting the firm to build your pool. Let us look at the specifications utilized in my pool. They include four major categories: size, design, construction, and equipment.

SIZE 14 to 21 feet in width and 32 feet long; 4 feet at the shallow end to 8 feet at the deepest.

DESIGN Free form (see sketch in Figure 6-2).

CONSTRUCTION Pneumatically placed monolithic Gunite®: 5 to 10 inches thick: Reinforced with ½- and ⅜-inch steel bars: Steel bars to be placed 12 inches both ways (vertically and horizontally) in the shallow end and 6 inches both ways in the deep end.

1. Steps: Gunite custom-built steps (three) in shallow end corner.
2. Tile: 6 by 6-inch blue trim around entire pool.
3. Coping: 8-inch brick around entire pool.
4. Deck: None. (We had the subcontractor who prepared our patio and foundation do the decking with the same wash gravel material. Their cost was much lower than the pool company's.)
5. Interior finish: Genuine pure white marble deluxe interior pool finish.
6. Other: Concrete filter slab to support tank and equipment.

EQUIPMENT

1. Filter System: Swimquip® model # MKW hi-rate diatomaceous with 25 square feet of filter area with stainless steel filter tank, 1 horsepower, 220 volt motor, hair and leaf strainer, face piping, fittings, suction valves, automatic air relief, and pressure gauge.
2. Heater: None. (We decided not to install a heater based on the fact that New Orleans weather permits swimming five to six months a year without heating the pool. In colder weather, we didn't care to be outside swimming. Provisions were made, however, so that a heater could be installed later. Space was left for it next to the filter system and a gas stub out was installed nearby. A pool heater may vary in price from $500 to $1,000 depending upon size, quality, and installation charges.)
3. Chlorinator: Custom-built tiled chloramatic feeder.
4. Diving Board: Deluxe 8-foot genuine Douglas fir, turquoise color coat with white sand, tread top surface, and stainless steel supports.
5. Ladder: None. (We felt that a ladder cluttered up a small pool and is a needless expense, especially if you have steps.)

6. Vacuum Cleaner: 10-inch vacuum cleaner with 30-foot floating hose and 16-foot handle. (There is a vacuum cleaner available that will automatically clean the floor and sides of a pool, but it is expensive—$300 to $400.)

7. Wall Brush: Deluxe 17-inch nylon brush with aluminum bracket and 16-foot handle.

8. Leaf Skimmer: Deluxe hand skimmer with a plastic screen and 12-foot handle.

9. Test Set: Dial-A-Test® set with case.

10. Thermometer: None. (If you buy a heater, the thermometer could be a useful item. The hand test suits my needs—if it's too cold, I don't go swimming.)

11. Safety Rope: None. (We installed the cycolac cup anchors. If necessary, the safety rope could be purchased later to separate the shallow end from the deep end.

12. Underwater Light: Chrome-plated 500 watt daylight-blue type complete with lens, bulb, deck box, and switch at filter area.

13. Time Clock: Automatic time clock for filtration equipment. (This enables you to regulate the time you wish the pool filter system to operate. It may vary from zero to 24 hours a day.)

14. Fill Spout: None. (You may wish to buy a chrome-plated brass one-inch fill spout; it is utilized to add water to the pool and would have been placed under our diving board. I once bumped my head on a spout at a neighbor's pool because it hangs over the edge. The injury potential and added cost convinced me that it would be better to install a water outlet near the pool and use a hose. This has proved satisfactory. I also use the hose to wash down the deck and water the shrubbery.)

15. Main Drain Fitting: 8-inch drain frame and grate.

16. Vacuum Wall Fitting: Vacuum fitting located in skimmer.

17. Return Water Fitting: Twin-jet returns with whirlpool action.

18. Skimmer: Deluxe super-flow self-adjusting type.

19. Other: 135 feet of 4-inch Orangeburg® pipe to be installed to storm sewer drain with three grills. Four gallons of muriatic acid, 40 pounds filter aid, 1 gallon tile cleaner, and 100 pounds of chlorine.

When to build

Begin building your pool after the pilings have been driven. At this point, there will be easy access to your entire lot. If you wait until your home is completed, it could be difficult for the pool builder to work with his equipment. In some situations, it has meant shoveling by hand and removing dirt in wheelbarrows. This can increase the cost of your job considerably.

Construction of a pool is messy work and you will have a real clean-up task if your home is already occupied and landscaped.

If you build your pool during the construction of your home, it is important to have some form of temporary protection, such as fencing to keep stray animals and youngsters out. You should also have necessary insurance to protect you from various types of injuries.

Building Phase

It is important to be present during construction so you can determine if the sketch and specifications are actually being implemented. The following aspects should be observed closely.

Design

Stakes will be placed to indicate the layout of the pool. It should conform to your drawing; check to see that it has the proper dimensions. Likewise, its location on your property should be in accordance with the plot plan. Review your city and subdivision requirements with respect to a swimming pool (see Figure 2-4).

Excavation

See that the excavation equipment does not damage sidewalks, neighboring property, or your building material. After the hole has been dug, measure the depth to determine if it complies with the specifications. Trucks will haul away the dirt; if you plan to build up your front yard, make certain that sufficient quantity is left to do so.

Preparation

The next step will be for workers to smooth out the bottom and edges so that it conforms precisely to the pool sketch. In my case, this required the services of three men for eight hours.

The main drain was then formed up (see Figure 6-3) and a load of shells placed at the bottom of the pool. Next, steel bars were put in the pool and held in place by ties.

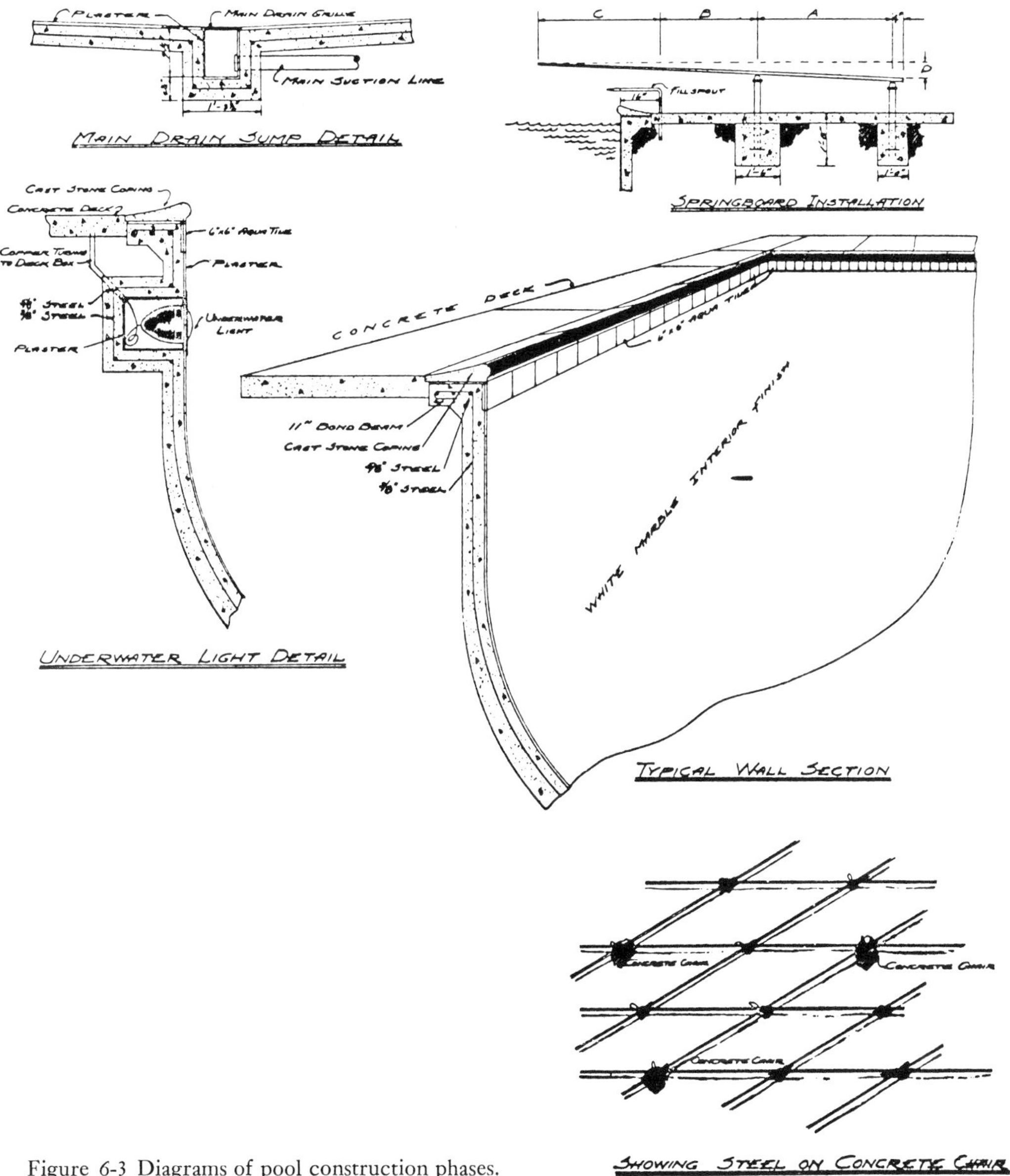

Figure 6-3 Diagrams of pool construction phases.

After pipe has been laid from the main drain to an outlet at the top of the pool, concrete is sprayed against the sides. This is referred to as a Gunite process.[5] It takes a powerful man to handle the pneumatic spray gun and it is

[5] You may desire to use another type of material for your pool such as metal, fiberglass, or vinyl. Check to find out which is best suited for your area and pocketbook.

interesting to watch this process. We checked to see that the floor had approximately 5 inches of concrete and the sides 10 inches. Other tasks included coping; tiling; and installing drainage, electrical equipment, and filtering system.

The plastering of the pool is an interesting job which results in a white marble interior finish. Once this material is dry, the pool should be filled with water. After the diving board is installed, decking can be completed for the pool area, driveway, and patio.

After your pool is finished, inspect and test the equipment. Be sure that there are no leaks and that final payment is withheld until the city inspectors have given their approval.

The Completed Pool

We wanted our pool to be seen from the den and located adjacent to the patio (Figure 6-4). We also wanted a diving board and made the depth 8 feet to minimize possibilities of injury.

Make certain your pool is deep enough to suit your needs. Our shallow end is 4 feet deep in order to allow good swimming in all areas. Wide steps were placed at the shallow end.

Figure 6-4 Free-form designed pool and surroundings.

The coping is a red brick. This color blends with our home and, in this area, is more reasonably priced. The decking is wash gravel and conforms with the driveway and patio.

Note the two doors leading to the garage. In one we store pool equipment —chemicals, diatomaceous earth, acid, and hose for the vacuum cleaner. The other door leads into a dressing room where the towels and swim suits are kept.

Maintenance

A swimming pool requires considerable care, but this care pays off, giving you an attractive pool and reducing repairs. I recommend that you, or a member of your family, clean your pool. It is good exercise and alerts you to any problems.

During the swimming season, you will need to add chlorine and acid frequently to maintain clear sparkling water. Check your acid and chlorine content daily, utilizing your test kit.

Chlorine and acid can be dangerous; handle them with care. The following warning for chlorine appears on the container:

> This material contains chlorine. When measuring any chlorine product, use only a scoop or container that is non-flammable and to be used *exclusively* for the bleach. Scoop or measure should be made of plastic or enamel or metal. Scoop must be completely dry, clean, and free of any foreign matter, liquid or solid. Chlorine is a strong oxidizing agent. Contact with heat, acids, greasy rags, paper, or any organic or oxidizing materials (such as vinegar, paint products, beverages, tobacco, cleaners, soap products, pine oil, kerosene, paper, rags, gasoline, mineral spirits) may cause fire. In case of fire, douse liberally with water.
>
> Never mix chlorine bleaches with anything except water. Avoid contact with the eyes.

Vacuum your pool weekly, clean the tile, skim the pool surface, backwash the filter, wash the deck, and empty the skimmer basket. It is desirable to have your pool inspected each year by a specialist.

A personal daily inspection is also important. Pranksters occasionally toss items over the fence. Eggs are difficult to clean up and the sooner removed the better. Toilet paper rolls can clog the filter system. Copper pennies, nails, and bobby pins cause rust marks if not removed promptly. If rust spots should occur, remove them with an abrasive compound.

Ask for advice on how to maintain the equipment and service the pool. You should also receive an instruction manual. Study it with care and go over the procedures several times with the pool representative. *Be sure you understand exactly how everything works.* Mistakes can be costly. For example, if you close off all the valves and turn on the motor, leaks will develop quickly.

You should be shown what to do when the filter system loses its prime—that is, doesn't have water in the pipes. Why call a repair man and pay $10 or more per visit when you can do it yourself?

Pool Safety

A pool is a delightful form of recreation, but it can be dangerous. Here are some rules which should be established:

1. Never swim alone.
2. Have a life line that can be thrown to a person. We use our garden hose.
3. Don't allow running or horseplay around the pool.
4. Wait at least an hour after eating before swimming.
5. If you have guests, check to see if they can swim. If not, use a safety rope to block off the deep end.

Summary

A swimming pool is a major investment. If you are considering a pool, your research should begin before the lot is purchased and when you initially plan your home. Select a lot that affords some privacy. You should spend adequate time and effort obtaining the facts prior to making your decision to build.

Prior to obtaining bids, it is desirable to prepare specifications and a sketch of your proposed pool. Select a first rate contractor who has a fine reputation and adequate financial resources.

It is normally less expensive to build a pool at the time you are constructing your home. If you decide to wait, interest charges for a loan may vary from 9½ percent to 42 percent.

A swimming pool is a continuing responsibility and requires considerable maintenance. It is preferable to do much of the work yourself; it saves money and, because of the personal interest, you normally do a better job.

CHAPTER 7 Protecting Your Property

The best time to consider appropriate security measures for your home is during the planning phase. This will result in lower costs and will provide immediate protection.

Crime

Unfortunately, most people do not provide adequate protection until a burglary has occurred. Crime is now the number one concern in our cities and the home is a prime target. Approximately $300 million in personal property was stolen last year; only $15 million was recovered. Families are disturbed not only by the rising number of burglaries but about their own safety.

This chapter tells what you can do to increase home security. You may refer to Figure 7-1 which lists protective measures and their costs, varying from zero to expensive.

A Managerial Approach

As a good manager, you should look at all aspects of a security system for your house before deciding what to purchase. Determine what will best meet your needs in relation to the money you can afford to spend. By including requirements in the initial planning of your home, the interest will be at the current first mortgage rate. If you decide to borrow to make such purchases later, the costs may range from 9½ percent to 42 percent.

A systems approach to security means considering each item in relation to your total home protective needs. Locks, for example, are worthless if you don't have a sound method of checking to be sure they are secured. Likewise, if you have a fixed amount of money available, you must establish priorities and decide what will best meet your requirements. Security, in turn, should be viewed as a subsystem of the entire house requirements. This means you

Items	*None*	*Cost* *Minor* *($1-$10)*	*Average* *($11-$100)*	*Expensive* *($101 up)*
Alarm system				
Alert to tricks	x			
At home appearance		x		
Dog		x	x	x
Escape routes	x			
Fence				x
Garage				x
Guns		x	x	x
Insurance				x
Iron doors and windows				x
Lighting			x	
Locks			x	x
Marking equipment		x		
Neighbors	x			
Pictures		x	x	
Police assistance	x			
Security checks	x			
Security during construction		x	x	x
Sitter service				x
Store valuables			x	
Top secret measures		x	x	x
Unpunctual hours	x			
Windows secured		x	x	
Wireless equipment				x

Figure 7-1 Protective measures and their expense.

should decide how much to allocate for security in relation to plumbing, electrical equipment, and all other expenses. If you wait to consider protective measures until the home is nearly complete, you hay have no money to provide such extras as quality locks. In building, one tends to do the essentials first and emergency expenses often use up any reserves.

Protective Measures

Here are some of the security measures that may be taken to protect your home.

Dogs

A good dog can be a valuable means of protection; a penetrating bark can scare off would-be intruders and can alert members of the family to take other precautionary measures such as calling the police, turning on the alarm system, or securing weapons. If you desire information about a trained watchdog, contact the canine division of your police department. (The police train dogs only for their own use or for other police departments. You might call a private guard agency if you are interested in having a trained watchdog.) These animals cost from $150 to $300, and two of the best are German shepherds and Doberman pinschers.

There is the risk, however, that a watchdog may turn on a member of the family. We had a friend who owned a boxer. It was extremely loyal to one of his daughters. For some unknown reason, it attacked him one day and lunged for his groin. It tore a gash in his leg and took off part of his finger. Fortunately, the daughter drove up at the same moment and called off the dog. I spoke to him after his return from the hospital. He commented; "If my daughter hadn't arrived, my life would have been in real danger."

Dogs can be quite expensive. The initial cost and maintenance for a small size thoroughbred in one community is as follows:

Initial Cost	
AKC registration	$ 5
License	6
Initial examination	8
Purchase price	125
	$144

Annual Cost	
Food	$ 96
Toys	10
Tags	6
Visits to veterinarian	60
	$172

It is possible to eliminate the purchase price by obtaining a dog from a friend or the Society for the Prevention of Cruelty to Animals, although the SPCA does require payment of $18 to cover the cost of shots and license.

A dog requires daily care. If you take a trip, it means placing him in a kennel or leaving him with a neighbor or relative. This eliminates your home protection during your absence.

Marking Equipment

Obtain an engraving tool that will enable you to mark your social security number or driver's license number on each piece of valuable equipment such as radio, television, or tape recorder.

Our community has an excellent procedure for identifying property. It is under jurisdiction of the police department and involves the following:

1. Fire stations loan out the marking equipment and provide an instruction sheet (Figure 7-2) and a form to list valuable items (Figure 7-3).
2. The police department is notified that the valuables have been appropriately marked (Figure 7-4), and they mail the homeowner two sets of decals.
3. The homeowner places the decals on his windows to inform all conerned that his equipment is marked.
4. The completed list of valuables (Figure 7-3) should be placed in a safe deposit box at the bank.

A number of cities have adopted a similar program. For further information, write the New Orleans Police Department in regard to "Operation Identification."

List of Valuable Items and Methods of Identification

Tape recorder	Radio
Television	Stereo
Air conditioner	Guns—pistol, rifle, shotgun
Camera	Typewriter and business machines
Hub caps of automobile	Wheels of automobile
Lawn mower	Tool box
Kitchen appliances	Repair and garden tools

All items should be marked with the *Louisiana drivers license* or *social security number* of the head of the home. The marking should be in a place of prominence where it can be easily observed without dismantling the object. This can easily be done on most items without detracting from their decorative appearance. Examples: On the chassis of the rear of a television set; on the inside of a stereo; on the inside or on the outer edge of automobile hub caps.

If an item such as a piece of jewelry cannot easily be marked, a complete description should be made and kept. These types of items should then be placed together on a light background and photographed. The owner should keep two sets of photographs, one which can be supplied to the police if a burglary occurs and the other for identification purposes. Valuable clothing such as fur coats and stoles should be marked with a felt pen in such a place that it would not destroy the attractiveness of the item but still be prominent when viewed by investigators.

Figure 7-2 Sample list of valuable items and methods of identification.

Valuable Items That a Burglar Would Most Likely Steal from My Residence

Item	Make	Year purchased	Serial number

After entering all valuable items on this form, place this form in a secure location in order that you may give investigating officers all pertinent data in case of a crime.

Figure 7-3 Sample form for listing valuable items.

Locks

There are a variety of locks for exterior doors to provide varying degrees of security. Let us classify them into four major categories.

KEY IN THE DOOR KNOB. This is the lock that a majority of homes use. It is the easiest to break into and the least expensive. In the construction of speculation homes, builders normally utilize this type.

Our carpenter showed us how quickly he could enter our home with this type of lock. With a twist of a wrench, he opened the door immediately. He also showed me how to gain entry by inserting the blade of a pocket knife between the latch and the door frame. This particular lock even had a trigger bolt which was supposed to prevent such action, but it didn't work.

The carpenter emphasized that the lock itself might be strong but is of little value unless the door has good hinges, is securely encased in a solid frame, and is made of a sturdy material. It takes little effort on the part of a burglar to break the panel of a weak door or knock off the latches and enter your home quickly.

DEAD LOCK. This lock is installed in the face of the door and can be obtained with or without a key. It requires drilling sizable holes in your door and frame. A dead lock normally has a rectangular-shaped metal projectile which is inserted into the hole of the frame to lock the door. The longer and

DATE

This is to verify to the New Orleans Police Department that I have affixed my Louisiana drivers license number to the valuables in my residence.

NAME

ADDRESS

TELEPHONE NUMBER

DRIVERS LICENSE NUMBER

I do not possess a drivers license, therefore, I have used by social security number to engrave the valuables in my residence.

NAME

ADDRESS

TELEPHONE NUMBER

SOCIAL SECURITY NUMBER
(If Used to Mark Items)

Two sets of decals will be mailed to the address placed on this form.

Figure 7-4 Sample form for informing police department of your valuables.

larger the projectile, the more difficult it is to pry the door apart. There is no beveled edge or spring mechanism on the dead lock so a pen knife or credit card cannot open it.

VERTICAL BOLT. This is an auxiliary lock which is screwed onto the door and frame. It precludes a burglar from jimmying the door because this type of lock has a pin which fits into a plate, as shown in Figure 7-5. We purchased vertical auxiliary locks for $11 each and find them most practical. Note the installation instructions for a Yale® lock (Figure 7-5). Unless you are qualified, I would suggest having a locksmith do it.

Yale® 197, 198, 197¼ and 198¼ Rim Deadlocks

INSTALLATION INSTRUCTIONS

FOR WOOD DOOR AND FRAME (DOOR 1⅜″ to 3″ THICK)

DOOR OPENING INWARD

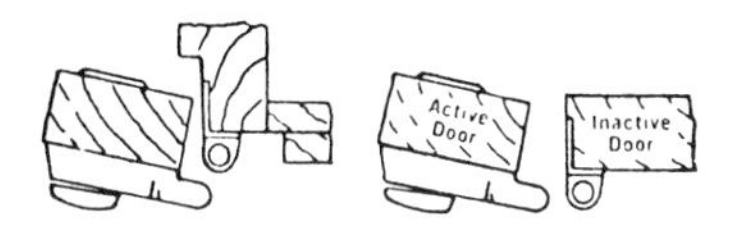

DOOR OPENING OUTWARD (And Sliding Doors)

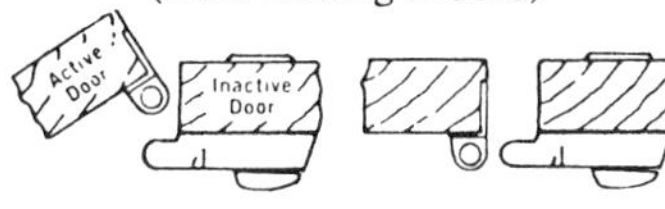

SINGLE DOOR DOUBLE DOORS DOUBLE DOORS SLIDING DOORS

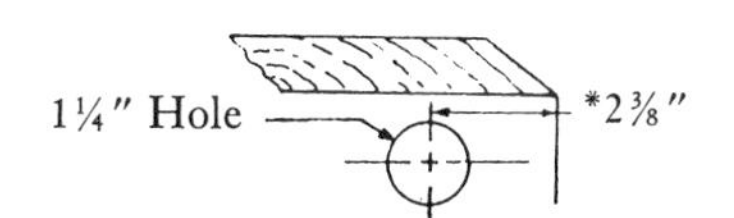

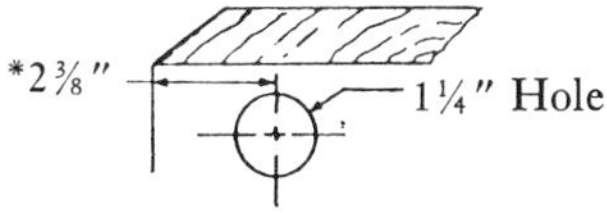

1 .... Measure 2⅜″ from inside edge of door and bore 1¼″ dia. hole. (Use care not to splinter door as drill breaks through).

*Note: If clearance between door and frame (or between double doors) exceeds ⅛″, locate 1¼″ dia. hole 2½″ from frame (or door).

2 .... INSTALL CYLINDER WITH YALE RIGHT SIDE UP

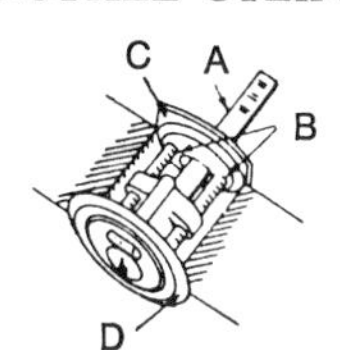

Slide cylinder ring *D* against cylinder head. Back plate *C* fastens cylinder to door. Break screws *B* to suit door thickness. Break bar *A* to project ⅜″ from face of door.

Withdraw key from cylinder so bar *A* will be in horizontal position.

3 .... BEFORE APPLYING CASE TO DOOR

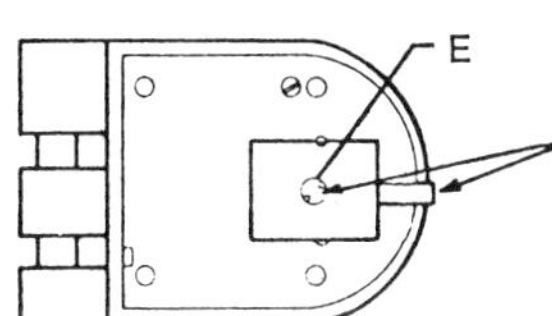

Hold shutter open, remove cardboard strip, and engage bar *A* in opening *E*.

Fasten lock case to face of door.

(One way screws are packed with 197¼ and 198¼ locks)

4 .... APPLY STRIKE

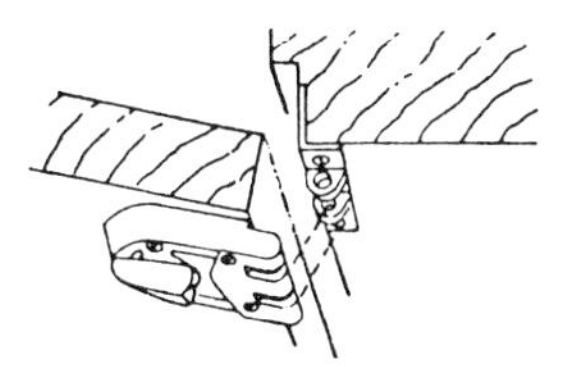

Align loops of strike with slots in nose of lock.

Caution: Be sure that strike and lock case are in vertical alignment, with top and bottom edges flush.

Base of strike flange (with 3 holes and loops) must be flush with surface of door. (May be mortise or surface applied.) Strike flange (with 4 holes) must be mortised.

Fasten strike with screws.

EATON Security Products & Systems Printed in U.S.A. 80-9710-0001-010 (4-72) Eaton Corporation Lock and Hardware Division

Figure 7-5 Sample locks and instructions for installation. (Courtesy of Eaton Corporation, Lock and Hardware Division)

Special Locks. You may wish to consider a lock with an alarm. If one tries to pick it or force entry, an alarm is sounded. There are pick-proof cylinder locks that make it difficult for the professional burglar to open. Bar locks can be purchased to install at the base and top of your doors. To lock them, metal projections are inserted in the ceiling and floor.

There is a wide choice of quality locks available today. I would suggest you visit several locksmiths and examine their merchandise. Also talk with security officers at several major companies prior to determining which is best suited to your needs.

Guns

A rifle or pistol can be an effective means of protection if you know how to shoot it accurately and can reach it on a moment's notice. Some homeowners keep a loaded weapon under their pillow or in a nearby table drawer, but this is very risky. A mistake could be fatal. There is an added danger with your children. In order to preclude their playing with weapons, you must put them in a safe place and separate ammunition from the guns. These extra safety measures, however, could mean that the guns would be unavailable if needed in a hurry.

If you do decide to buy a weapon, register it with the police. Check with them to see if it is possible to receive instructions on how to use it properly. Pistols, in particular, require considerable training to be an expert shot. Practice must be continued to keep up your expertise. Perhaps there is a local gun club available that you might wish to join. Your weapon should be cleaned periodically so that it functions when needed.

Weapons are very dangerous and may cause more harm than good. I was a Regular Army Infantry Officer for 23 years and I have had several close calls as a result of guns being "accidentally fired."

Neighbors

It helps to be friendly with your neighbors. Get to know them and work out some means of mutual cooperation. If you are away, they can look after your property. You can be helpful to them, for example, by reporting to the police any suspicious person on their property.

By working as a team, much more can be accomplished than by individual effort. You may find it desirable to organize a group of homeowners to patrol the neighborhood. If necessary, you may feel a need to hire a private organization to cope with the problem. To be effective, any type of arrangement should be worked out with the support and agreement of most neighborhood families.

This good neighbor policy, however, can be overdone, as recently happened in a nearby residential area:

> One evening a newly arrived neighbor noted a "stranger" carrying furniture to an unmarked vehicle from the house next door. She called the police. They arrived after the truck departed but one of the officers knocked on the door of the home reported as being robbed. A lady opened it and the policeman explained that they had received a call on a suspicious-looking individual who was taking equipment out of her house. The lady responded: "What do you mean! That's my husband. He is moving out. We are getting a divorce."

Police Assistance

The police can be your friend. Make a point to do your part by contacting the police department and asking their views on security measures. They may send an officer to your home to offer advice. You may also wish to have a policeman speak on home protection at a neighborhood meeting. He can provide excellent precautions that apply to your specific locality. It will also give you an opportunity to ask questions. At a recent meeting, a speaker gave worthwhile advice on safeguarding your home against criminals and tips for vacationers (see Figure 7-6).

By showing an interest in the police, they will be more alert to your problems. Cooperation is a two way street. You may find it helpful to ask them to patrol your area more frequently. In addition to the city protection, your locality may have their own police force. If so, contact this group.

Call the police when you see any suspicious person, strange vehicle, or unusual action in your neighborhood. The following loss might not have occurred if the police had presented their views on security measures:

> A neighbor observed a man loading two television sets on his truck. They were being taken from the house of a young working couple. She thought they were going to have their television sets repaired and proceeded to walk across the street and asked: "Are you repairing the Smiths' television sets?" His reply: "Yes Ma'am." So she said: "Mine doesn't work. Can you take it along and make the necessary repairs? I would appreciate it." He was most obliging: "I won't charge you for pick-up service and will bring it back next Tuesday." A week passed and the television set had not been returned. That evening she waited until her neighbors arrived home from work and then asked: "Have

Safeguard your home against criminals

KEEP outside lights on all night to illuminate all sides of your home.
MAKE sure that the area between the home and detached garage is lighted.
STORE expensive jewelry in a safe deposit box.
IF you leave the house, close the garage door. An open garage door is an open invitation to the daytime burglar.
LEAVE lights on in the house when you are away for the evening.
BEFORE you leave for a vacation, inform the police and a neighbor. Leave window shades and blinds at a normal level and make sure that milk bottles and newspapers do not accumulate at your door. Ask post office to hold mail.
DON'T give keys to strange service men. Separate auto keys and house keys. Check references of maids and cleaning people. Keys are easily duplicated.
LOCK windows and doors–including the side doors–when you retire. Use a supplementary chain lock.
LOCK attic windows. Lock garage too.
PUT a strong short chain on your door.
DON'T leave your door unlocked, even when you're in the house.
KEEP the door chained when a stranger knocks until you are satisfied his purpose is legitimate.
DOUBLE-LOCK your door when you leave.
ALWAYS test your doors to make sure you have locked them before leaving.
SEE that shrubbery that might provide an easy hiding place is kept trimmed.
CHECK thoroughly the references of maids or service people who are given access to your home.
MAKE sure all screens are secured from the inside.
A DOG provides warning and protection.

Tips for Vacationers

STOP all deliveries such as newspapers, milk, and laundry. Ask a neighbor to remove mail from mailbox daily and remove circulars and brochures left at the house.

LOCK all windows and doors securely.

USE a light meter which turns on a lamp and outside lights at dusk.

IF absent for an extended period, arrange to have your lawn mowed, preferably by a relative or trusted neighbor.

NOTIFY police that you will be on vacation, giving address, date and time of departure, and date and time of return.

MARK personal belongings with your driver's license number.

IMPORTANT PHONE NUMBERS

____
____
____
____
____
____
____
____

Figure 7-6 Advice for safeguarding your home.

you gotten your television sets back from the repair shop?" Mr. Smith replied: "What do you mean? We don't have any television sets. Our house was robbed last week and they took both of them."

Alarm System

A burglar alarm system can provide both psychological and physical protection. A potential burglar can see that outside your front door is an on-off lock device with a red lamp and a warning bell housed in a tamper-proof box. Decals may also appear in your window.

The physical aspect is the system itself. In the event of robbery, an alarm is set off once entry is made by means of a door or window. The theory is that the noise will frighten the burglar away and would give you time to obtain a weapon. Emergency switches can also be installed in various locations to sound the alarm if you see something suspicious.

The house alarm system, in some communities, can be connected directly to police headquarters. In other localities, the home can be tied into the alarm company's office which in turn would contact the police. For this central hookup, there is a monthly charge.

The alarm system decision should be made in the planning phase of home-building. It is more reasonable to have it installed during construction, and this gives you protection from the moment you move in.

Alarm systems are expensive. They may vary from $700 to $1,200. It is important to make certain that you are dealing with a reputable firm that will back up its equipment. Prepare your specifications and be precise in your requirements. If possible, obtain three bids based on *free* estimates. These should be submitted after a study is made of your house plans and alarm system specifications.

You should also give consideration to an effective fire alarm system. One type has a fire alarm circuit which when broken by fire sounds a warning horn. It is also possible to install a unit that detects smoke or heat and activates an alarm. By having the combination of heat, smoke, and fire detection, you are in a better position to be warned in time. But such a system is expensive. If you have a multiple-story home, an adequate fire alarm system would appear to be a more important requirement.

Enclosed Garage

An enclosed garage is a good place to keep your bicycle, car, and other equipment. Our son recently left his bike at the front door step one Sunday afternoon. He went inside for several minutes and upon returning it was gone —not to be found again.

Install necessary locks for each garage door and make daily checks to be sure that they are secured. An automatic electric garage door is a convenience, saving you from locking and unlocking the door each time you enter or leave with your car.

Iron Doors and Windows

Today, more homeowners are installing iron grills at their entryways and windows. In some cases, this motif can be attractive; however, if used at all windows and entrances, it can make a house look like a prison. This grill work is expensive and requires good locks to be effective.

This does provide added security since it means another barrier to penetrate before reaching the interior of the house. But if you put an iron door at your front entry only, the home is still vulnerable at other entrances and windows. A burglar will normally seek to enter a home at its weakest location. This is why it is so important to consider all aspects of security in devising an effective means of protection.

Other Suggestions

STAGGERED HOURS. Try not to establish a precise pattern for leaving or returning home so that people watching your residence will find it difficult to make break-in plans.

Robbers much prefer to work an unoccupied house where there is less possibility of being identified or engaged in a fight in which they might be caught, injured, or killed.

AT-HOME APPEARANCE. You may wish to turn on your record player, television, or radio before leaving the house. This gives the impression that someone is at home. At all times, give your home an occupied appearance.

STORE VALUABLES. Don't keep your valuables at home. If you have heirlooms or other expensive items, keep them in a safe deposit box at the bank.

PICTURES. It is helpful to take pictures of your possessions. Keep one set in the safe deposit box. Be sure to up-date your photographs as you acquire additional items. Your memory can play tricks but your camera will provide solid proof.

Escape Routes. You should devise a plan to leave your home rapidly in the event of a burglary or fire. Check your fire department for suggested avenues of escape and what to do in case of fire. I have found them most helpful and they will come to your home to recommend fire prevention measures. This will include proposals for equipment needed such as fire extinguishers.

Windows Secured. You may want to nail or screw your windows shut. This makes it more difficult for someone to use this type of entry. However, in the event of fire, it could make escape difficult from certain locations —particularly if you have a multiple story residence. You might want a screwdriver, hammer, or chair handy in the event it is necessary to open the window.

Outdoor Lighting. A good outdoor lighting system can be helpful. You can obtain lights that come on at dusk and go off at dawn. There is one negative aspect to this system: it illuminates your home and could draw it to the attention of a person who might otherwise have ignored it.

Enclosed Fence. A fence can be another barrier to burglars gaining entry to your home. Its value is enhanced if the gates are kept securely locked. A fence, however, could also be a hiding place. Once intruders are in the backyard, they could work unobserved. Shrubbery could also be used as a hiding place if too profuse.

Wireless Equipment. There are radar devices that can be purchased to alert you of potential burglars. This equipment can be placed in strategic locations such as by doors and windows. When a person passes within a certain distance, it sounds an alarm. An advantage over a wired alarm system is that the alarm goes off before the house is broken into. A problem with this device is that a friend, animal, or a high wind could trigger the alarm.

In contrast to the motion detection system, there is also a wireless monitor system. Small radio transmitters are placed on windows and doors. A master control is also installed; and if an entry is made, this device sounds an alarm and automatically alerts the alarm company's office which in turn calls the police. Installation costs may vary from $300 to $800. A monthly service charge begins at $15.

Alert to Tricks. The following anecdote points out the need to watch for trickery that may result in your home being robbed:

> While a lady was walking downtown, she had her purse snatched. The next day she received a telephone call at home. The person said, "I

> found your billfold this morning while walking to the office. It has no money in it, but contains your credit cards and driver's license."
>
> The lady was overjoyed and promised a reward for the return of these items. He said, "I don't have a car, but you could meet me downtown at the Namllits Drug Store. I could take a few minutes off from work to be there at 3 o'clock this afternoon." He described himself well so that there would be no question of mistaken identity.
>
> That afternoon the lady went to Namllits at the appointed hour. She waited 45 minutes but no one showed up. Upon returning home, she found that her home was robbed. On reporting it to the police, she was informed this trick had been pulled the previous week in another section of town.

Other ruses to gain entry include posing as a policeman, FBI agent, repairman, or mailman. A fake accident and request for help may lure you from your home. It pays to be alert.

Burglars read about obituaries, wedding ceremonies, and graduations and note when the occasion will be held. People who announce their vacation plans in the newspaper or elsewhere also provide good information to robbers.

SECURITY CHECKS. Establish a good security check system. It doesn't pay to install quality devices and not use them. You can buy the best locks, but if you don't keep them secured, they serve no purpose. Establish a workable procedure, for example, to ensure that your alarm system is turned on and off as required. One home has a sign at each exterior door reminding members of the family to lock it. Another household makes one person responsible to see that all doors and windows are locked before bedtime.

TOP SECRET MEASURES. The Watergate hearings before a Senate Committee, in 1973, revealed sophisticated devices used by CIA and FBI trained personnel. The potential security measures you can develop are limited only by your imagination and availability of money. These items might include hidden cameras, recording machines, peepholes, eavesdropping devices to pick up conversations on your property, monitoring your phone, fingerprinting entry ways, trip wires, and other booby-traps. The more unknowns you have, the more protection you'll have against a successful burglary.

Fortress Concept

You can turn your house into a fort with an alarm system, iron doors and window guards, electronic devices, special locks, watchdogs, elaborate lighting systems, and an arsenal of weapons inside. But is this the way you want to live? Keep in mind that if someone really wants to enter, he will. No fortress in history has proved to be invulnerable.

In the long run, you can improve your home security by being concerned about crime in our community, state, and nation. The fortress concept of home security only means that the burglar will turn to a less well-protected residence. The home security race, like the arms race, is not the most desirable solution.

Insurance

It is important to acquire appropriate insurance with a reputable, first rate company. Your homeowner's policy may provide you with the needed coverage; read it carefully to be sure.

A homeowner's policy insures your home against fire, lightning, wind, hail, and many other perils. I recommend the broadest coverage to include "all physical loss."

Comprehensive personal liability insurance (CPL) is included in the homeowner's policy. This coverage provides protection should you or any member of your household be held legally liable for actions resulting in accidental injuries to others or accidental damage to the property of others. Coverage applies to personal and sports activities, and it is effective both on and away from the premises and inside and outside the residence.

The following are some situations in which the CPL insurance would be effective:

1. A deliveryman or a guest is injured in your residence.
2. You accidentally injure someone while engaged in golf, hunting, or horseback riding.
3. One of your children injures a playmate with one of his toys.
4. Your pet bites someone.
5. A fire spreads from your property to the property of others.

The special medical-payment clause and the property-damage clause, which form an integral part of the CPL insurance, do not necessarily require legal liability to be effective. The insured's obligation may be moral rather than legal. Payment under these clauses may be authorized by the insurance company without resorting to a court trial.

Be sure to obtain flood insurance if you are in one of the 213 flood-prone American communities in 31 states. People in these areas may purchase insurance at federally subsidized premium rates. The maximum amounts range from $17,500 for single-family homes up to $30,000 for two to four-family dwellings. Rates vary from 40 to 50 cents per $100 of insurance, depending on the value of the structure. Contents also may be insured up to $5,000 with rates of 50 to 60 cents per $100 of insurance. This insurance protection is available through local agencies. A local agent reports that on a $17,500 single home, the premium would be $70 per year. To obtain $5,000 in personal-property coverage would cost $25.

The National Flood Insurance Act of 1968 states that "anyone who does not choose to buy flood insurance in a community which is eligible will not be able to get federal financial disaster aid for flood losses that occur after one year from the date his community became eligible to the extent that the loss would have been covered by flood insurance."

The rising cost to replace possessions makes it important for you to make annual reappraisals to be sure that you have adequate protection in the event of fire, burglary, or other loss.

During the construction of your home, take out appropriate insurance. If an individual is working directly for you, then have coverage in the event of his injury. Also make certain that your subcontractors are properly insured.

Security During Construction

The following measures can be taken to keep losses at a minimum while building your home.

1. Enclose your garage promptly so that it can serve as a storage area for tools and other materials.

2. Build your home rapidly. The less idle time the better.

3. Order out only the amount of supplies needed to accomplish the immediate jobs.

4. Ask your neighbors to keep an eye on the premises.

5. Pay for protection. In one area, much of the vandalism was done by youngsters living in the neighborhood. They would turn over paint cans and write obscene words on doors and walls. Windows were broken, pipes bent, and materials taken. To combat this theft, a builder asked one of the young toughs to "protect" his property.

A house under construction is a fascinating place for children and draws them like a magnet. You can expect some damage unless there is 24-hour supervision. One builder in our area hires a night watchman who comes on duty as the subcontractors are leaving each evening. He found his losses from vandalism were so great that it was cheaper to have a person protect his construction sites on a continuing basis.

6. Make surprise visits during the weekend and evening hours.
7. Enclose your house as soon as possible. Prior to installation of your permanent doors, your carpenter can nail up plywood sheets each evening.
8. Watch tools. It is important to alert workmen to protect their tools while they are on the job. In some locations, carpenters cannot leave their hammers or saws unguarded for a moment. For example, when I was helping to install the air conditioning unit, my job was to rivet the ducts together. It was 120 degrees in the attic when I decided to take a break. When I returned, the riveting gun was missing. I never recovered it, but fortunately my insurance company paid for it.

Summary

There are a variety of protective measures that can be taken to provide greater security for your home. Some, like alarm systems and iron doors, are expensive. In contrast, other actions can be taken for little or no cost. These might include neighborhood assistance programs, working with the police, marking valuable items, installing locks yourself, and taking security precautions.

The best time to determine your security requirements is during the planning phase of building your home. You can then allocate a specified amount of money for protective devices within your total budget. If you wait, installation costs will be higher, interest charges greater, and you will not have the desired protection from the moment you move in.

You should have adequate protection to give you a sense of security and peace of mind. This may require enough equipment to discourage an intruder and make him decide it would be easier to go elsewhere. You can help this decision by utilizing the devices at your disposal. Remember that security is only as good as you make it. If you fail to lock your doors or turn on your alarm system, these measures are a wasted expense.

Crime is on the increase in the United States. One approach to protection is for each home to be a fortress, but this is not a healthy, long-term solution. We can reduce the need for home security measures in the future by participating in local, state, or national programs to improve the quality of life in America.

CHAPTER 8
How to Keep Your Cool While Moving

Once your home is available for occupancy, you should plan to move in as quickly as possible. The slightest delay increases the chance of vandalism, results in extra expenses for living costs, and makes it impossible to check the workmanship and equipment on a 24-hour basis.

With adequate planning, you can make this move into your new home efficiently and economically. The fact that you are subcontracting gives you good control over the completion date and ample time to prepare for a successful move. Your PERT chart and timetable will enable you to determine when the house will be ready and you can make definite arrangements with the moving company of your choice. This move should be taken into consideration during the planning phase of your home.

If you have made previous moves you have, no doubt, discovered that each new change of location makes the task of packing, transporting, and unpacking your possessions more difficult. The accumulation of possessions can become a mountainous problem with the passing of time. After nineteen moves, I have come to respect my wife's motto: "If in doubt, pitch it out." But by pitch it out, she doesn't mean to throw it away. For years, the military thrift shops have been her happy selling grounds, and there are civilian parallels. Magazines are given to the hospitals and other items are donated to the Goodwill Industries. The result is that we never had an attic full of dust collectors.

Our numerous moving experiences in the military and academic worlds have varied from outstanding to disastrous. We sometimes relocated only across the street, and sometimes 10,000 miles. Twice we placed our possessions in storage and once we physically moved it all ourselves. Major carriers, government trucks, local carriers, and our family car have all been used at various times.

We learned from each move. Good management is of utmost importance. The closer the supervision, the better the move. There is no substitute for personal interest, but this requires a good deal of hard work on your part. For example, when we moved across the street while living in Arlington, Virginia, we had no loss whatsoever; and our expenses were zero. But to move 8,000 pounds took three of us two full days. We also had help from several neighbors. Obviously, such short distance moves are uncommon. But in each instance where we were unable to adequately supervise the packing, loading, and storage, we had poor results.

Management Process

The management process presented in the first chapter can be applied successfully to your next move. Let us list each aspect of this process and then discuss them as they pertain to moving.

1. Determine your objective.
2. Plan your move with care.
3. Implement your plan.
4. Check to see if the move is accomplished without damage and within the agreed price.
5. As the "move" manager, make sound decisions by defining your objective; wisely utilizing your resources (men, money, materials); and performing the functions of planning, organizing, and controlling within the framework of an effective organization.

WHAT	Move
WHEN	Upon completion of home
WHERE	From city X to city Y
HOW	Major van line
WHY	Accomplish it in an economical and efficient manner

Figure 8-1 The move objective should provide answers to the questions of what, when, where, how, and why.

Objective

Let us assume that you decide your objective is to move your household goods from city X to city Y upon completion of your home, hiring a major van line to accomplish the shipment in an economical and efficient manner.

As you note, the objective provides answers to the questions of what, when, where, how, and why. Figure 8-1 presents these answers in tabular form. Keep in mind that the move objective must be viewed as a sub-objective of the entire building project.

Planning Your Move

Adequate planning is absolutely essential for a successful move. Here are some points to consider:

Reduce Household Possessions

During the planning and building of your new home, make periodic examinations of your household possessions and be ruthless in discarding items you "may someday use," but that you know you never will. Those that have value can be disposed of by advertisement, garage sale, or through a thrift shop. The Salvation Army and Goodwill Industries are delighted to receive a call and will pick up your merchandise.

The advantages of keeping your household possessions to a minimum are twofold: (1) it can reduce your shipping costs markedly; (2) it saves considerable time and effort on your part in preparing for the move and arranging it in your new home.

You should also consider selling your furnishings at your present location, taking only valuable items with you. If you offer to sell your home with the drapes, rugs, basic furniture, and major appliances, it can be a more attractive deal to some people. Likewise, if you live in an apartment or rent a house, it may be possible for you to sell it furnished. I know of several instances where owners have sold their condominiums furnished and made a healthy profit. Many people like the idea of taking over a tastefully furnished residence and moving in with no concern about decorating.

If you decide to reduce your household possessions, it will also enable you to obtain a fresh motif for your new home. Perhaps the old furnishings would not do justice to the new surroundings. Prior to making a decision, weigh the various factors to include the cost of buying new furniture.

Select the Best Method

Use a moving company that has an excellent reputation. This can be checked by asking neighbors and friends who recently used the services of a mover. If possible, obtain estimates of costs from three competent firms in order to have a basis for comparison. If you plan to store some of your household goods, visit the warehouse. Consider the alternative of hiring a van and doing the packing and moving yourself. This can be a major undertaking, but for intra-city moves or short hauls, it can be accomplished at a savings in money, time, and breakage.

Provide Adequate Funds

In determining your budget requirements, be sure to allow yourself adequate *move-in* funds. By move-in funds I mean both moving costs (storage and moving) as well as expenses for new furnishings and appliances required at the time you first enter your new home. If you have to spend time enroute, also make an allowance for your own family travel expenses.

There are companies who pay for your entire move and will also guarantee that you do not take a loss in selling your old home. It is great if you have this good fortune, but your move can still be more agreeable if you follow sound management concepts.

Estimate Moving Expenses

You should be able to estimate your moving costs with reasonable accuracy. Each carrier you contact will send an estimator to your home to make an *estimate* of the total charge involved. Be sure to show him everything you plan to send by van so his bid will be accurate. Likewise, let him know what extra furnishings you may purchase prior to moving. Well in advance of your move, you should obtain a copy of the "Estimated Cost of Services." Note in Figure 8-2 the various services that determine the cost of a move: transportation; valuation charge; pickup and delivery for storage in transit; storage in transit; warehouse handling; special servicing of appliances; hoisting, lowering, or carrying pianos and heavy articles; packing and unpacking services; purchase of barrels, boxes, cartons, mattress covers, crates, and containers.

After studying the table of measurements (Figure 8-2), you may wish to come up with your own estimated total weight. Such an undertaking serves as a check on the estimator and may encourage you to eliminate some of your possessions by sale, donation, and the trash pile.

Mayflower **ESTIMATED COST OF SERVICES**

REFER TO THIS NUMBER IN ALL COMMUNICATIONS
Order for Service No.

NAME OF CARRIER **AERO MAYFLOWER TRANSIT COMPANY, INC.**

ADDRESS OF CARRIER **P.O. Box 107, Indianapolis, Indiana 46206 (317) 639-2451** DATE ____ 19__

NAME OF SHIPPER ____ ADDRESS ____ PHONE ____

SHIPMENT MOVING FROM ____ TO ____

SHIPPER'S DESTINATION CONTACT ____ PHONE ____ PACKING DATE OR PERIOD OF TIME REQUESTED ____

IMPORTANT NOTICE: This estimate covers only the articles and services listed. It is not a guarantee that the actual charges will not exceed the amount of the estimate. Common carriers are required by law to collect transportation and other incidental charges computed on the basis of rates shown in their lawfully published tariffs, regardless of prior rate quotations or estimates made by the carrier or its agents. Exact charges for loading, transporting, and unloading are based upon the weight of the goods transported, and such charges may not be determined prior to the time the goods are loaded on the van and weighed. Charges for additional services will be added to the transportation charges.

LOADING DATE OR PERIOD OF TIME REQUESTED ____

DELIVERY DATE OR PERIOD OF TIME REQUESTED ____

TOTAL ESTIMATED COST $____

I HEREBY ACKNOWLEDGE that I have received from (check one)
____the carrier supplying this estimate
____a carrier supplying another estimate
____other source
Summary of Information for Shippers of Household Goods, Form BOp 103

Signature of Shipper or his Representative

If the total tariff charges for the listed articles and services exceed this estimate by more than ten percent, then, upon your request, the carrier must relinquish possession of your shipment upon delivery in advance of the payment of the total amount of tariff charges shown on the bill of lading or freight bill. You are still obligated to pay the balance of the total charges within 15 days. Maximum amount to be paid on delivery of your C.O.D. shipment in cash, certified check or money order is (total estimated cost plus 10 percent):
$____

ESTIMATED COST OF SERVICE (Based on tariff ____ MF-I.C.C. NO. ____) — ESTIMATED CHARGES

Transportation: Est. wt. ____ lbs.; ____ mi.; @ $____ per 100 lbs. $____

Valuation Charge {for liability on part of carrier in excess of / that assumed when its lowest rates are charged}

On Transportation: $____ @ 50¢ per $100, or fraction thereof ____

On Storage-in-Transit @ ____¢ per cwt. (10% of monthly storage rate) for each 30 days or fraction thereof ____

Additional transportation charges: (Explain) ____

Pickup or delivery for storage in transit ____ lbs. @ ____¢ per 100 lbs. ____

Storage in transit at ____ lbs. @ ____¢ per 100 lbs. {for each 30 days or fraction thereof} ____

Warehouse handling ____ lbs. @ ____¢ per 100 lbs.(one time charge) ____

Extra pickup or delivery at ____

Special servicing of appliances ____

Hoisting, lowering, or carrying pianos, heavy articles ____ (Explain) ____

CONTAINERS (see below) ____

Packing (see below) ____

Unpacking (see below) ____

Labor ____ man/men for ____ hrs.; @ ____ (per man per hour) ____

Other services ____ (Explain) ____

COST ESTIMATE ☐ WAS ☐ WAS NOT GIVEN TO SHIPPER — TOTAL ESTIMATED COST: $____

ESTIMATED COST OF CONTAINERS, AND PACKING AND UNPACKING SERVICES	CONTAINERS			PACKING			UNPACKING		
	Estimated Number	Per Each	TOTAL	Estimated Number	Per Each	TOTAL	Estimated Number	Per Each	TOTAL
BARREL, dish-pack, drum, etc.		$	$		$	$		$	$
BOXES, not over 5 cubic feet									
Over 5 not over 8 cubic feet									
CARTONS: Less Than 1½ cubic feet									
1½ cubic feet									
3 cubic feet									
4½ cubic feet									
6 cubic feet									
6½ cubic feet									
Wardrobe Carton									
Crib Mattress Carton									
Mattress Carton, Regular (Not exceeding 54"x75")									
Mattress Carton, King Size (Exceeding 54"x75")									
Mattress Cover (plastic or paper)									
CRATES AND CONTAINERS: (Specially designed for mirrors, paintings, glass or marble tops and similar fragile articles) Gross measurement of crate or container									
	Estimated Container	COSTS	$	Estimated Packing	COSTS	$	Estimated Unpacking	COSTS	$

Remarks ____

NOTICE: It is mandatory that the total cubic footage shown on the table of measurements be multiplied by not less than 7 to determine the total estimated weight. Articles not to be shipped should be indicated by a "check mark" in the column provided on the table of measurements.
If the prospective shipper has not previously been furnished with the Summary of Information For Shippers of Household Goods as required by the Interstate Commerce Commission, he should be furnished at this time. (Table of measurements on reverse side)

AGENCY ____ SIGNATURE AND TITLE OF ESTIMATOR ____

Figure 8-2 Sample form from reliable movers for estimated cost of services. (Courtesy of Aero Mayflower Transit Company, Inc.)

TABLE OF MEASUREMENTS

Article A	Article B	Article C	Article	Cubic feet per piece	No. of pieces	Cubic feet
			LIVING AND FAMILY ROOMS			
			Bar, Portable	15		
			Bench, Fireside or Piano	5		
			Bookcase	20		
			Bookshelves, Sectional	5		
			Chair, Arm	10		
			Chair, Occasional	15		
			Chair, Overstuffed	25		
			Chair, Rocker	12		
			Chair, Straight	5		
			Clock, Grandfather	20		
			Day Bed	30		
			Desk, Small or Winthrop	22		
			Desk, Secretary	35		
			Fireplace Equipment	5		
			Foot Stool	2		
			Lamp, Floor or Pole	3		
			Magazine Rack	2		
			Music Cabinet	10		
			Piano, Baby, Gr. or Upr.	70		
			Piano, Parlor Grand	80		
			Piano, Spinet	60		
			Radio, Table	2		
			Record Player Port.	2		
			Rugs, Large Roll or Pad	10		
			Rugs, Small Roll or Pad	3		
			Sofa, 2 Cushions	35		
			Sofa, 3 Cushions	50		
			Sofa, 4 Cushions	60		
			Sofa, Sect., per Sect.	30		
			Stud, Couch or Hideabed	50		
			Tables, Dropl'f or Occas.	12		
			Tables, Coffee, End/Nest	5		
			Telephone Stand & Chair	5		
			Television Combination	25		
			TV. or Radio Console	15		
			Television Table Model	10		
			DINING ROOM			
			Bench, Harvest	10		
			Buffet	30		
			Cabinet, Corner	20		
			Cabinet, China	25		
			Chair, Dining	5		
			Server	15		
			Table, Dining	30		
			Tea Cart	10		
			Rugs, Large or Pad	10		
			Rugs, Small or Pad	3		
			BEDROOM			
			Bed, incl. Spring & Mattr:			
			Bed, Double	60		
			Bed, King Size	70		
			Bed, Single or Hollywood	40		
			Bed, Rollaway	20		
			Bed, Bunk (set of 2)	70		
			Bookshelves, Sectional	5		
			Bureau, Dresser, Chest of Dr'w'rs, Chifrb. or Chifnr.	25		
			Cedar Chest	15		
			SUB. TOTAL COLUMN 1			

Article A	Article B	Article C	Article	Cubic feet per piece	No. of pieces	Cubic feet
			BEDROOM			
			Chair, Boudoir	10		
			Chair, Straight or Rocker	5		
			Chaise Lounge	25		
			Desk, Small or Winthrop	22		
			Dresser or Vanity Bench	3		
			Dresser Doub. (Mr. Mrs.)	50		
			Night Table	5		
			Rug, Large or Pad	10		
			Rug, Small or Pad	3		
			Vanity Dresser	20		
			Wardrobe, Small	20		
			Wardrobe, Large	40		
			NURSERY			
			Bathinette	5		
			Bed, Youth	30		
			Chair, Child's	3		
			Chair, High	5		
			Chest	12		
			Chest, Toy	5		
			Crib, Baby	10		
			Table, Child's	5		
			Pen, Play	10		
			Rug, Large or Pad	10		
			Rug, Small or Pad	3		
			KITCHEN			
			Breakfast, Suite Chairs	5		
			Breakfast Table	10		
			Chair, High	5		
			Ironing Board	2		
			Kitchen Cabinet	30		
			Roaster	5		
			Serving Cart	15		
			Stool	3		
			Table	5		
			Utlity Cabinet	10		
			Vegetable Bin	3		
			APPLIANCES (Large)			
			Air Conditioner, Window	30		
			Dehumidifier	10		
			Dishwasher	20		
			Dryer, Electric or Gas	25		
			Freezer: (Cu. Capacity)			
			10 or less	30		
			11 to 15	45		
			16 and over	60		
			Ironer or Mangle	12		
			Range, Electric or Gas	30		
			Refrigerator (Cu. Capacity)			
			6 cu. ft. or less	30		
			7 to 10 cu. ft.	45		
			11 cu. ft. and over	60		
			Vacuum Cleaner	5		
			Washing Machine	25		
			PORCH, OUTDOOR FURNITURE & EQUIPMENT			
			Barbecue or Port. Grill	10		
			Bath, Bird	5		
			Chairs, Lawn	5		
			SUB. TOTAL COLUMN 2			

Article A	Article B	Article C	Article	Cubic feet per piece	No. of pieces	Cubic feet
			PORCH, OUTDOOR FURNITURE & EQUIPMENT			
			Chairs, Porch	10		
			Clothes Line	5		
			Clothes Dryer Rack	5		
			Garden Hose and Tools	10		
			Glider or Settee	20		
			Ladder, Extension	10		
			Lawn Mower (Hand)	5		
			Lawn Mower (Power)	15		
			Lawn Mower (Riding)	35		
			Leaf Sweeper	5		
			Outdoor Child's Slide	10		
			Outdoor Child's Gym	20		
			Outdoor Drying Racks	5		
			Outdoor Swings	30		
			Picnic Table	20		
			Picnic Bench	5		
			Porch Chair	10		
			Rocker, Swing	15		
			Roller, Lawn	15		
			Rug, Large	7		
			Rug, Small	3		
			Sand Box	10		
			Settee	20		
			Spreader	1		
			Table	10		
			Umbrella	5		
			Wheel Barrow	8		
			MISCELLANEOUS			
			Ash or Trash Can	7		
			Basket (Clothes)	5		
			Bicycle	10		
			Bird Cage & Stand	5		
			Card Table	1		
			Cabinet, Filing	20		
			Carriage, Baby	20		
			Chairs, Folding	1		
			Clothes, Hamper	5		
			Cot, Folding	10		
			Desk, Office	30		
			Fan	5		
			Fernery or Plant Stands	10		
			Foot Lockers	5		
			Garbage Cans	7		
			Golf Bag	2		
			Heater, Gas or Electric	5		
			Incinerator	10		
			Metal Shelves	5		
			Ping Pong Table	20		
			Pool Table	40		
			Power Tools	20		
			Sled	2		
			Step Ladder	5		
			Suitcase	5		
			Table, Utility	5		
			Tackle Box	1		
			Tool Chest	10		
			Tricycle	5		
			Vacuum Cleaner	5		
			Wagon, Child's	5		
			Waste Paper Basket	2		
			Work Bench	20		
			Sewing Mach. Portable	5		
			Sewing Mach. Cabinet	10		
			SUB. TOTAL COLUMN 3			

Article A	Article B	Article C	Article	Cubic feet per piece	No. of pieces	Cubic feet
			OTHER ITEMS			
			CONTAINERS (To Be Packed by Shipper)			
			Barrels	10		
			Boxes, Wooden	3		
			Boxes, Wooden	5		
			Boxes, Wooden	10		
			Boxes, Wooden	15		
			Boxes, Wooden	20		
			Carton			
			Less than 1½ cu. ft.			
			1½ cu. ft.			
			3 cu. ft.			
			4½ cu. ft.			
			6 cu. ft.			
			6½ cu. ft.			
			Wardrobe Furnished by Carrier	15		
			CONTAINERS (To Be Packed by Carrier)			
			Barrels	10		
			Boxes, Wooden	3		
			Boxes, Wooden	5		
			Boxes, Wooden	10		
			Boxes, Wooden	15		
			Boxes, Wooden	20		
			Carton			
			Less than 1½ cu. ft.			
			1½ cu. ft.			
			3 cu. ft.			
			4½ cu. ft.			
			6 cu. ft.			
			6½ cu. ft.			
			Wardrobe Furnished by Carrier	15		
			SUB. TOTAL COLUMN 4			
			TOTAL COL. 1			
			TOTAL COL. 2			
			TOTAL COL. 3			
			TOTAL COL. 4			
			GRAND TOTAL			

SUMMARY

______ CU. FT. @ ______ LBS. PER CU. FT. ______ LBS.

ESTIMATED TOTAL WEIGHT ____________ **LBS.**

CODES: ABC A — ARTICLES LOADED B — ARTICLES UN-LOADED C — ARTICLES NOT TO BE SHIPPED

Form R-42 (Rev. 7/72)

(Estimated Cost of Services on Reverse Side)

Figure 8-2 Sample form from reliable movers for estimated cost of services.

SUMMARY OF CONTENTS VALUES

Classification	Present Value
Living Room	
Dining Room	
Library or Den	
Bedrooms	
Recreation Room	
Bathrooms	
Musical Instruments	
Furs	
Kitchen, Pantry, Etc.	
Laundry	
Table and Bed Linens	
Porch and Lawn	
Garage and Miscellaneous	
Nursery	
Basement or Utility Room	
Clothing	
Personal Belongings	
Jewelry	
Silverware	
TOTAL	

Mayflower WORLD-WIDE MOVING

America's Most Recommended Mover

[illegible] SERVICE CORP.
[illegible]—PACKING
[illegible] World-Wide Service
846 BARONNE ST. — MAgnolia [illegible]53
NEW ORLEANS, LOUISIANA

F-27

LITHO IN U. S. A.

Name________________

YOUR PERSONAL

Inventory

of

HOUSEHOLD FURNISHINGS AND PERSONAL PROPERTY

Date__________

In order to protect your household goods sufficiently under the Mayflower Comprehensive Protection Plan, when you move long-distance, it is necessary to know their value accurately. Even though you are not moving long-distance, and therefore do not require this Plan, it is still vitally important for you to have a record of your possessions and their value. It is the only way you can know what insurance coverage you need, and the best way to establish claims in case of loss. We hope this form will be helpful in compiling this record.

AERO MAYFLOWER TRANSIT COMPANY, INC.
INDIANAPOLIS, INDIANA

Figure 8-3 Inventory of household furnishings and personal property. (Courtesy of Aero Mayflower Transit Company, Inc.)

LIVING ROOM

Article	How Many	Year Bought	Cost	Present Value
Chairs				
Tables				
Davenport				
Rugs				
Carpet				
Lamps				
Pictures				
Mirrors				
Piano				
Organ				
Radio				
Clock				
Curtains				
Draperies				
Fireplace Fittings				
Vases				
TV Set				
Record Player				
Records				
Hi-Fi				
Air Conditioner				
TOTAL				

DINING ROOM

Article	How Many	Year Bought	Cost	Present Value
Table				
Chairs				
China Cabinet				
Buffet				
Server				
Tea Cart				
Rugs				
Carpet and Pad				
Curtains				
Draperies				
Pictures				
Mirrors				
Dinner Sets				
China				
Glassware				
TOTAL				

LIBRARY OR DEN

Article	How Many	Year Bought	Cost	Present Value
Chairs				
Tables				
Desks				
Davenport				
Rugs				
Carpet				
Lamps				
Pictures				
Radio				
Record Player				
Mirrors				
Clock				
Bookcases				
Books				
Curtains				
Draperies				
TV Set				
Typewriter				
Files				
TOTAL				

BEDROOMS

Article	How Many	Year Bought	Cost	Present Value
Beds and Springs				
Mattresses				
Chest				
Chairs				
Vanity				
Dressing Tables				
Clocks				
Rugs				
Carpets				
Lamps				
Pictures				
Bedside Tables				
Toilet Sets				
Curtains				
Draperies				
Sewing Stands				
Cedar Chest				
Radios				
TV Sets				
TOTAL				

RECREATION ROOM

Article	How Many	Year Bought	Cost	Present Value
Chairs				
Tables				
Rugs				
Pictures				
Billiard Table				
Ping Pong Table				
Card Tables				
Curtains				
Books				
Radio				
Lamps				
TV Set				
Bar				
Hi-Fi				
Games				
TOTAL				

BATHROOMS

Article	How Many	Year Bought	Cost	Present Value
Cabinet				
Chairs				
Scales				
Clothes Hamper				
Heaters				
TOTAL				

MUSICAL INSTRUMENTS

Article	How Many	Year Bought	Cost	Present Value
Violin				
Mandolin				
Banjo				
TOTAL				

FURS

Article	How Many	Year Bought	Cost	Present Value
Coats and Capes				
Neckpieces				
TOTAL				

KITCHEN, PANTRY, ETC.

Article	How Many	Year Bought	Cost	Present Value
Refrigerator				
Stove				
Deep Freeze				
Kitchen Cabinet				
Floor Covering				
Curtains				
Chairs				
Tables				
Utensils				
Dishes				
Supplies				
Radios				
Dishwasher				
TOTAL				

LAUNDRY

Article	How Many	Year Bought	Cost	Present Value
Washer				
Ironer				
Tables				
Chairs				
Electric Irons				
Dryer				
Dehumidifier				
TOTAL				

TABLE AND BED LINENS

Article	How Many	Year Bought	Cost	Present Value
Sheets				
Pillow Cases				
Blankets				
Spreads				
Table Cloths				
Napkins				
Luncheon Sets				
Towels				
Wash Cloths				
Bath Mats				
TOTAL				

Figure 8-3 Inventory of household furnishings and personal property.

PORCH AND LAWN

Article	How Many	Year Bought	Cost	Present Value
Chairs				
Tables				
Couch				
Curtains				
Glider				
Swing				
Gym Set				
Rotisserie				
TOTAL				

GARAGE AND MISCELLANEOUS

Article	How Many	Year Bought	Cost	Present Value
Bicycles				
Lawn Mower				
Garden Tools				
Garden Hose				
Vacuum Sweeper				
Sewing Machine				
Hand Tools				
TOTAL				

NURSERY

Article	How Many	Year Bought	Cost	Present Value
Crib				
Chairs				
loor Covering				
Chest				
Toys				
TOTAL				

BASEMENT OR UTILITY ROOM

Article	How Many	Year Bought	Cost	Present Value
Work Bench				
Hand Tools				
Power Tools				
Machinery				
Canned Goods				
TOTAL				

CLOTHING

Article	How Many	Year Bought	Cost	Present Value
Men's Suits				
Men's Coats				
Women's Dresses				
Women's Coats				
Children's				
Hats				
Shoes & Boots				
TOTAL				

PERSONAL BELONGINGS

Article	How Many	Year Bought	Cost	Present Value
Golf Equipment				
Tennis Rackets				
Stamp, Coin Collec.				
Fishing Tackle				
Guns				
Luggage				
TOTAL				

JEWELRY

Article	How Many	Year Bought	Cost	Present Value
Watches				
Rings				
Necklaces				
Bracelets				
Brooches				
TOTAL				

SILVERWARE

Article	How Many	Year Bought	Cost	Present Value
Tea Set				
Trays				
Knives				
Forks				
Spoons				
Holloware				
TOTAL				

CHECKLIST

- ☐ Change address, get records, trip arrangements.
- ☐ Collect and sort mail for listing address changes.
- ☐ Notify post office of move and fill out change of address cards.
- ☐ Send address change to friends and businesses.
- ☐ Get all medical and dental records.
- ☐ Check and clear tax assessments.
- ☐ Have your W-2's and other tax forms forwarded.
- ☐ Transfer insurance records, check auto licensing requirements.
- ☐ Notify school and make arrangements for sending transcripts of school records to new school.
- ☐ Have letters of introduction written.
- ☐ Arrange for transfer of jewelry and important documents.
- ☐ Close charge accounts.
- ☐ Arrange shipment of pets and any immunization records.
- ☐ Make travel plans.
- ☐ Get hotel reservations and make note to reconfirm.

14 DAYS BEFORE YOUR MOVE

- ☐ Collect all clothing or items to be cleaned or repaired.
- ☐ Return things borrowed—collect things loaned.
- ☐ Have bank transfer accounts and release safe deposit box.
- ☐ Arrange to disconnect utility services.
- ☐ Arrange to connect utility service at new home.
- ☐ Have farewell parties and visits.
- ☐ Make arrangements to have heavy appliances serviced for move.
- ☐ Give away articles you don't plan to take along, give to charitable organizations, get signed receipt for tax purposes.

7 DAYS BEFORE YOUR MOVE

- ☐ Dispose of all flammables.
- ☐ Have car inspected and serviced.
- ☐ Pack suitcases ahead of time.
- ☐ Select traveling games.
- ☐ Set things aside to pack in car.
- ☐ If you haven't made arrangements with your mover to do so, take down curtains, rods, shelves, television antenna if agreement with new owner authorizes this.
- ☐ Start packing suitcases you can live out of, if necessary, for the first day in your new home.
- ☐ Line up a baby sitter for moving day so you can look after moving.
- ☐ In a special carton, place items you will need in the first few hours in your new home: soap, towels, coffee, cooking pot, etc. Mark this carton with sticker "Load Last—Unload First!"
- ☐ Make up special cartons with "Do not move" for articles to be taken in car.

DAY BEFORE MOVING

- ☐ Empty and defrost your refrigerator and freezer and let them air at least 24 hours. Also clean and air your range.

- ☐ Line up a simple breakfast for next morning that won't require refrigeration or much cooking. Use paper plates.
- ☐ Finish packing personal belongings, but leave out the alarm clock!
- ☐ Get a good night's rest.

MOVING DAY

- ☐ Be on hand the day of your move, or have someone there authorized to answer questions.
- ☐ Accompany the van operator while he inventories your possessions to be moved.
- ☐ Make last minute check on your appliances to see that they have been serviced.
- ☐ Sign (and save your copy) of bills of lading and make sure delivery address and place to locate you enroute are correct.
- ☐ Advise driver exactly how to get to new residence.
- ☐ Also delivery date or dates.
- ☐ Ask that you be advised of final cost. (This will be determined after van is weighed.) Then, if you have not arranged for time payment of move, or your company is not paying for it, make sure you'll have the needed cash, money order or certified check to pay before van is unloaded at destination because carriers require payment before unloading.
- ☐ Strip your beds, but leave fitted bottom sheets on your mattresses.
- ☐ Before leaving house, check each room and closet, make sure windows are down and lights out.

MOVING-IN TIPS

- ☐ Upon arrival at new location, call the Mayflower agent immediately to leave address and phone number where you can be reached and when, so you can make final arrangements for delivery.
- ☐ Be on hand at unloading and have a plan for placement of your furniture.
- ☐ Check all electrical fuses. Sometimes pennies have been used as substitutes!
- ☐ Check the condition of your belongings. If any items are missed or damaged, note this on your inventory sheet and shipping papers; then report such information to your Mayflower agent who will take care of it for you.
- ☐ If your utilities haven't been connected, call them for this service, and have them check your appliances for proper operation.

Figure 8-4 Checklist of arrangements to be made when moving. (*Moving Kit*, Copyright 1970, Aero Mayflower Transit Company, Inc.)

Storage Considerations

If it is necessary to put your furniture in storage, check with care the warehouse you use. Find out exactly where your furniture will be placed and, if possible, look at it in storage prior to your departure.

You can make storage arrangements direct with a local warehouse or secure an agreement with a major carrier. If you have an agreement with an interstate van line, the terms and conditions of your contract apply to the storage as well as the move. In contrast, if you use a local warehouse, the terms are whatever you arrange with that firm.

Normally the carrier will set a time limit on this "temporary storage" of no more than 180 days. After that date, it goes into "permanent storage" and the warehouse rules apply. This is another reason for completing your home promptly. The longer your possessions remain in storage, the greater the risk of damage or loss.

Record Possessions

Keep an up-to-date record of your possessions. This should include the original cost, year purchased, and the estimate of present value. If losses are sustained, you will then have appropriate evidence to support your claim. This list can also be helpful in the event of household burglaries as discussed in the previous chapter. Most large moving companies can supply forms for this purpose. See Figure 8-3 for an example of an inventory of household furnishings and personal property.

Use a Checklist

Give consideration to taking actions as indicated in Figure 8-4 at both your present and new address. Well in advance of your move, study this checklist with care. There are some excellent suggestions and you may wish to act on them before and during your move. Without some effective reminder, it is so easy to forget something important in the busy days before your departure.

Carry Valuables

If possible, take your valuable possessions and heirlooms with you. Be sure to have appropriate insurance. Some people also carry bedding rolls, kitchen utensils, and other necessary items in their car so if the shipment arrives late they will have adequate furnishings to sleep in their home.

Have Adequate Loss Protection

How much protection should you have on your shipment?[1] If you decide not to have additional protection for your household possessions, in the event of damage, you will only be paid at a rate of 60 cents per pound per article. For example, let's assume you have a lightweight bicycle that weighs 10

[1] In this instance, we are referring only to shipments by motor common carriers of household goods operating in interstate and foreign commerce. If you are making an intrastate move by a local carrier, you should determine his protection.

pounds, cost you $250, and was purchased just prior to shipment. If it had been totally damaged during shipment, you would receive $6 and suffer a loss of $244. You can, however, pay an extra amount of money for greater protection. You have two choices: you can determine the value of your entire shipment and be covered for this amount in the event of a complete loss, or the alternative is not to set a valuation. In this case, the mover's maximum liability is automatically set at $1.25 times the weight of your shipment in pounds. If we assume that the net weight totals 5,000 pounds then you would receive $6,250 in the event of a total loss.

The bill of lading reads as follows with respect to the valuation:

> Unless the shipper expressly releases the shipment to a value of 60 cents per pound per article, the carrier's maximum liability for loss and damage shall be either the lump sum value declared by the shipper or an amount equal to $1.25 for each pound of weight in the shipment, whichever is greater. The shipment will move subject to the rules and conditions of the carrier's tariff. Shipper hereby releases the entire shipment to a value not exceeding
>
> ______________________________
> (to be completed by the person signing below)
>
> NOTICE: The shipper signing this Request For Service must insert in the space above, in his own handwriting, either his declaration of the actual value of the shipment, or the words "60 cents per pound per article." Otherwise, the shipment will be deemed released to a maximum value equal to $1.25 times the weight of the shipment in pounds.

If your merchandise is actually lost enroute, you receive a cash payment based on the protection you selected. If there is damage, however, the carrier has the choice of either restoring the article to the same condition when it was taken from your home or paying you the actual value less depreciation.

Keep in mind that your coverage is not called insurance and the Interstate Commerce Commission actually prevents movers from selling insurance. The ICC calls your protection a matter that affects "the liability of the mover for loss or damage to your goods."

Read the ICC Pamphlet

The Interstate Commerce Commission (ICC) has prepared a pamphlet entitled "Summary of Information for Shippers of Household Goods." Every prospective shipper should receive a copy. The ICC has made it mandatory for all carriers to provide their customers with copies. However, an ICC study indicates that about 18 percent of the people who shipped their household goods did not receive a copy.

This booklet lists the following do's and don'ts and I urge you to read its entire contents well in advance of your move:

Do

Read this information booklet *entirely*.
Select your household goods mover with care.
Be sure that agreements between you and the carrier are in writing and on the order for service and the bill of lading.
Examine and make sure that physical inventory record of your household goods is accurate as to number of items, condition of furniture, and so on.
Make sure you understand the limited liability of the household goods carrier.
Schedule your departure and arrival with enough flexibility to allow for possible failure on the part of the carrier to meet exactly his scheduled time.
Accompany, if you can, your carrier to the weighing station for weighing of your shipment.
Advise the carrier of a telephone number and/or address where you can be reached enroute or at destination.
Request a reweigh of your household goods if you have any reason to believe the weighing is not accurate.
Be certain that everything on the inventory is accounted for *before* the van operator leaves either origin or destination.
File a claim for any loss or damage noted on the delivery papers as soon as possible.

DON'T

Fail to read this information booklet entirely.
Believe an estimate is a final cost of your move.
Expect the carrier to provide boxes, cartons, barrels, or other packing material free of charge.
Expect maid service and appliance service free of charge.
Plan to leave your old residence until the moving company leaves, unless you have a friend or neighbor acting on your behalf.
Fail to make arrangements to have in cash or certified check the maximum amount shown on the order for service unless credit has been arranged for in advance.
Expect your household goods to be unloaded until you have paid at least the maximum amount shown on the order for service in cash or certified check unless credit has been arranged for in advance.

Sign any receipt for your household goods until you are certain that they are all delivered and that there has been no apparent damage that has not been noted on the shipping papers.

The ICC has 82 offices in major cities throughout the United States. Each office employs a minimum of two people who are available to answer your questions and provide other assistance in regard to their area of responsibility.

Implementing Your Move Plan

Arrange for their Arrival

During the packing and moving phase, it is important for you or a member of your family to be present. Plan for the packers' arrival. Have your household possessions neatly arranged and provide adequate working space for the workers. Items to be packed should be clean and readily available.

Observe the packers' work carefully. You can learn much from professionals on how best to protect your china and crystal for a safe move. Some other time you may decide to do it yourself.

You may wish to code your containers. By having them clearly marked, it will help you to place them in the correct room at your new home. Should any breakage occur while your items are being packed, make a notation on the bill of lading. Likewise if the packers indicate that your furniture is gouged, scratched, or otherwise damaged and you disagree, state your disagreement in writing. This written statement will be essential on submission of your claim.

Inspect with Care

Once the movers start taking furnishings from your residence, you should have a system of checking each item that is placed in the van. This check also gives you an opportunity to see if other furnishings are already aboard. Unless you are able to fill a van, it will normally take on other household possessions enroute in order to have a full load. This may delay your shipment.

Your personal supervision will normally result in movers taking greater care. It won't take long to judge if you have a good crew. If they are sloppy and hurrying the job, inform the leader. If it continues, contact the local office immediately to send a responsible supervisor.

Weigh Your Possessions

Before the moving van leaves, you should receive a bill of lading. This is your receipt for your household possessions and will indicate the weight of the van prior to placing your merchandise on it (tare weight). You may wish to follow the van to the weighing station to determine the weight of your load. The weight with your possessions on the van is called the gross weight. Your charge is arrived at by subtracting the tare weight from the gross weight (net weight). Once the net weight has been established, you can call the carrier's local office and obtain the transportation cost of your shipment.

If you have any concern about the accuracy of the weight of your household possessions, you can request that they be weighed again before they're unloaded. But in the event there is not a certain variation in the weight, you will be required to pay a reasonable charge for this second weighing.

Checking On Your Move

Unloading Inspection

The control aspect of your move occurs when the truck arrives and you actually have an opportunity to check each item. Take time to inspect your possessions with care.

Try to be at your new home prior to the arrival of the van. If you keep it waiting, you may find that your shipment has been placed in storage. This can be expensive. Find out how long a van will wait prior to leaving your residence.

Once the van arrives, it helps to have a sketch available so your furnishings can be properly placed in the various rooms. It makes movers unhappy when you frequently switch furnishings. One family I know takes photographs of their possessions. After a move, they find that a look at the photos enables them to resettle more quickly. The pictures also serve as a fine record.

You should expect the carrier to furnish qualified people. We have on one occasion found a driver who picked up a derelict as a helper. We reported it, but only after several items were damaged and it was apparent the helper was drunk.

Check each item as it comes off the van to see if it is damaged. Make appropriate notations on the inventory, delivery receipt, or bill of lading.

If your agreement calls for unpacking, then be sure that all items are removed from the cartons and placed in the various closets, cabinets, and drawers that you designate. Again make notation of any damage found.

Payment

The movers will be anxious to complete unloading, have you sign for the merchandise, and receive payment. Payment for your shipment should not be made until you are satisfied all work has been completed in accordance with the contract. Make certain that you have listed all damaged and missing items and it is so stated above your signature. *Don't be rushed.*

Unless you have been given credit by the mover prior to delivery, payment must be made in cash, certified check, money order, traveler's check, or cashier's check.

Submitting a Claim

After rechecking for damaged and missing items, submit your claim promptly and follow up to see that you get a fair deal. The ICC states in the previously mentioned pamphlet:

> If you need to file a claim, the earlier this is done, the quicker the mover can make settlement. We cannot stress enough that your best proof of claim is notation on the bill of lading, inventory, or delivery receipt at the time of delivery. If you should later discover that an article was lost or damaged and you have proof that such loss or damage was caused by the mover, you can still file a claim for such loss within 9 months after the move. Your claim is much more difficult to process if it is delayed or presented some time after your goods have been delivered.

There is a splendid article in the May 1973 issue of *Consumer Reports* entitled "Moving? Still Lots of Potholes Along the Way." The variation in claim settlements by major carriers is shown in Figure 8-5.

Short Distance Move

We have discussed the long distance move, but what about short hauls? If you build your home in the community you presently live in, you have additional options in deciding how to make such a move. Your choices include: employing a large carrier; using a local firm that specializes in short-haul moving; renting a truck and driver; renting the truck and driving it yourself; moving the possessions yourself, using family transportation; or a combination of the aforementioned.

If you have the time and energy, there is no question that you can save considerable money and do a splendid job by taking on much of the responsibility yourself. Over the years, we found our most successful moves were those

The big carriers' performance records

This chart shows the 20 biggest household-goods movers' performance records in some important consumer areas for the last six months of 1972. The information was obtained from reports the companies themselves prepared and submitted to the ICC in accordance with that agency's regulations. Companies are listed alphabetically.

	Claims refused	*Claims closed in 30 days or less*	*Claims taking over 120 days to close*	*Frequency of under-estimates*
AERO MAYFLOWER TRANSIT CO.	10%	57%	13%	21%
ALLIED VAN LINES	3	43	10	24
AMERICAN RED BALL TRANSIT CO.	15	27	20	23
ATLAS VAN LINES	13	15	27	24
BEKINS VAN LINES	6	64	2	25
BURNHAM VAN SERVICE	39	78	0	35
FERNSTROM STORAGE AND VAN CO.	17	58	0	24
GLOBAL VAN LINES	11	38	26	18
GREYHOUND VAN LINES	14	38	15	29
JOHN F. IVORY STORAGE CO.	12	66	12	22
KING VAN LINES	8	57	0	15
LYON VAN LINES	11	70	11	29
NATIONAL VAN LINES	20	54	4	35
NEPTUNE WORLD WIDE MOVING	20	58	3	16
NORTH AMERICAN VAN LINES	28	38	13	23
REPUBLIC VAN AND STORAGE CO.	31	27	0	23
TRANS-AMERICAN VAN SERVICE	34	13	24	8
UNITED VAN LINES	13	22	17	21
U.S. VAN LINES	17	71	2	21
WHEATON VAN LINES	7	42	7	23
Average for all 20 carriers	17%	47%	10%	23%

Figure 8-5 Performance records of the major household-goods movers. (Copyright 1973 by Consumers Union of United States, Inc., Mount Vernon, New York 10550. Reprinted by permission from *Consumer Reports*, May 1973.

we did ourselves. But it is hard work. The fact that you are building your own home permits you to move in items once you think the house is well secured. But a word of caution. It may be wise not to move many items before you arrive. There is always the possibility of pilferage and theft.

Your automobile can be useful for moving small items. You can then rent a truck and perhaps hire a college student who will help you move.

Obtain bids from local movers and speak to friends who have used their services. The fact that you will be there to supervise is a major factor in making a successful move.

Summary

To simplify your move and reduce costs, it is desirable to sell or give away furnishings that will not be utilized later. Be ruthless in eliminating possessions that won't be appropriate for your new home, as well as dust collectors that have not been used in years.

Have adequate protection for your household possessions while in transit so that in event of loss or damage you will be fully covered. If possible, take your valuables and heirlooms with you; under no circumstances should you ship them in a regular move.

Move into your new home as rapidly as possible after its completion. Utilize a management approach in making your move. Determine your objective, plan your move with care, implement it, and check to see that the move is completed as planned. Performing these duties requires supervision—the better the supervision, the more successful the move.

CHAPTER 9
Landscaping

Shortly after arriving at your new home, you should begin landscaping—Webster defines landscaping as changing "the natural features of a plot of ground so as to make it more attractive, as by adding lawns, trees, bushes."

Throughout the book, we've emphasized the planning phase. If you have done your planning properly, it should include landscaping. This will permit you to implement the plan when you move in.

What Type of Landscaping Is Appropriate?

Landscaping gives you an opportunity to be creative: it should reflect your personality and be appealing to you. However, you must take into consideration the possibility of resale, the time available for maintenance, costs involved, and future needs based upon the changing composition of your family. If you have a large family, or numerous other interests, you probably will not be able to spend much time keeping the grounds attractive. Simplicity may be your primary concern. This could be accomplished with considerable concrete; artificial grass; and a minimum of shrubs, flowers, and trees. Taking this course, however, will not endear you to the growing number of environmentally conscious citizens.

Budget for Landscaping

During the planning stage, it is important to budget an adequate amount of money for landscaping. This is where families often make short cuts. They either fail to consider it or decide to wait until they are well settled. You can do this, of course, but your property will be unsightly. Good planning will allow you to make your residence attractive from the start.

How Do You Go About Landscaping Your Home?

There are several approaches you may wish to consider in landscaping your home. Here are some of the possibilities.

Advice from the County Agent and Land Grant University

County agents can be found in every state. They work with a land grant university and are paid in part from funds provided for by the Department of Agriculture. If a county agent is not nearby, you can write your land grant university. In Louisiana write to: Director, Cooperative Extension Service, Louisiana State University and Agricultural and Mechanical College, Baton Rouge, Louisiana. Some of the free publications available are entitled: *Enjoy Your Home Grounds*, *Diseases and Insects of Lawn Grasses*, *Shrubs for Louisiana Landscapes*, *Plan Landscapes for the Family*, and *Roses.*

The university will also test the soil in your yard to determine what may be needed to improve it (see Figure 9-1).

Secure a Gardener

Talk with neighbors who have attractive yards and find out who is taking care of them. Select the best qualified individual to tend to your grass, shrubs, and flowers.

Hire a Professional Landscaper

Contact a professional landscaper and have him prepare a design for your house and give you an estimate of what the entire job will cost. You should try to obtain three bids.

Do it Yourself

In order to keep your costs at a minimum, it may be necessary to do it yourself. Landscaping your own property, particularly if you have a small lot, is not a difficult task. You can get splendid ideas from reading available literature and observing attractively landscaped homes.

A neighbor went to a professional landscaper initially but could not afford his price. He also obtained an estimate from a local gardener. He then figured what it would cost if he did it himself, including buying materials and equipment. He found that it was considerably cheaper to use his own labor. Today he has one of the most attractive yards in the neighborhood.

SOIL TESTING LABORATORY

L. S. U. Agricultural Experiment Station, Baton Rouge 3, Louisiana

NAME	POST OFFICE	DATE OF SAMPLING	LABORATORY NUMBER	SOIL TYPE	SAMPLE IDENTIFICATION
Dr. Richard Stillman	New Orleans, La. 2311 Oriole St.	9-68	10821	sl	

ANALYSIS OF SOIL

		VERY LOW	LOW	MEDIUM	HIGH
Available Phosphorus, (parts per million*)	61				✓
Available Potassium, (parts per million)	110			✓	
Available Calcium, (parts per million)	4000+				✓
Available Magnesium, (parts per million)	380				✓
pH (acidity or alkalinity)	8.7	alkaline			
Organic Matter, (per cent)					
Soluble Salts, (per cent)	0.019	✓			

*Parts per million means "pounds per million pounds of soil." An acre of soil 3 ½ inches deep weighs about one million pounds.

RECOMMENDATIONS: - (These are the least amounts of fertilizer that you should use per acre.)

CROP	LIME NEEDED (TONS)	TOTAL NUTRIENTS NEEDED PER ACRE: NITROGEN (N)	PHOSPHATE (P_2O_5)	POTASH (K_2O)	SUGGESTED FERTILIZER TO MEET THE RECOMMENDATION SEE BACK OF THIS SHEET FOR MORE INFORMATION
					______ lbs. of ______ and ______ lbs. of nitrogen, or use equivalent amounts of other grades or materials. SEE OTHER SIDE.
					______ lbs. of ______ and ______ lbs. of nitrogen, or use equivalent amounts of other grades or materials. SEE OTHER SIDE.
					______ lbs. of ______ and ______ lbs. of nitrogen, or use equivalent amounts of other grades or materials. SEE OTHER SIDE.
					______ lbs. of ______ and ______ lbs. of nitrogen, or use equivalent amounts of other grades or materials. SEE OTHER SIDE.

ADDITIONAL REMARKS: The salt is not harmful. However, the pH is too high, apply sulfur at the rate of 4 ounces per square yard of surface area and mix well with the soil to lower ~~low~~ pH.

Figure 9-1 Sample form from soil testing laboratory.

A New Lawn

If possible, your lawn should be established shortly after you move in. This ground cover is important. In the event of heavy rain, it can keep your soil from being washed away. Besides, it looks nice. Initially, you must determine which grass is most appropriate for your area. In New Orleans we looked at three types: St. Augustine, Zoysia Matrella, and Bermuda. We selected St. Augustine after discussing it with the county agent and observing the successes and failures of our neighbors. It provides a thick carpet of grass and, in our locality, gives an attractive appearance all year 'round.

INFORMATION ON FERTILIZERS

A complete fertilizer, such at 8-8-8, contains a certain number of pounds of total nitrogen, available phosphate and available potash.

FOR EXAMPLE:	TOTAL NITROGEN (N)	AVAILABLE PHOSPHATE (P_2O_5)	AVAILABLE POTASH (K_2O)
100 lbs. of 8-8-8 contains	8	8	8
100 lbs. of 6-12-6 contains	6	12	6
100 lbs. of 3-12-12 contains	3	12	12

WHEN YOU BUY FERTILIZER- consider the plant food in the bag as well as the price.

Our recommendation lists the minimum grade of fertilizer. You can substitute the following **grades** and **amounts** of fertilizer for those recommended, and obtain the same amounts of plant food.

Instead of using 100 lbs. of these recommended grades,	You could substitute **these amounts** of these fertilizers
8-8-8	80 lbs. of 10-10-10, or 67 lbs. of 12-12-12 or 62 lbs. of 13-13-13
6-12-6	60 lbs. of 10-20-10 or 50 lbs. of 12-24-12
3-12-12	60 lbs. of 5-20-20, or 50 lbs. of 6-24-24
0-14-14	70 lbs. of 0-20-20
5-10-10	83 lbs. of 6-12-12 or 50 lbs. of 10-20-20
5-20-10	72 lbs. of 7-28-14

WHEN YOU NEED NITROGEN, here are the main carriers of nitrogen and their nitrogen contents:

	LBS. NITRATE NITROGEN	LBS. AMMONIA NITROGEN
100 lbs. of anhydrous ammonia contains	--------------	82
100 lbs. of urea contains	--------------	45
100 lbs. of ammonium nitrate contains	16 ½	16 ½
100 lbs. of "uran" 32 contains	8	24
100 lbs. of ammonium sulfate contains	--------------	20
100 lbs. of cyanamid contains	--------------	21
100 lbs. of nitrate of soda contains	16	--------------

When soils are kept properly limed, each pound of nitrogen from each of these sources gives about the same increase in yield. You should **always** keep your soil limed properly, regardless of the source of nitrogen you use.

Figure 9-1 Sample form from soil testing laboratory.

We decided to build our lawn four to five inches above our walk and bought three loads of top soil for this purpose. Instead of doing the job ourselves, we hired a gardener who had done quality work in the neighborhood. He purchased the grass, shrubs, and fertilizer. Next time, we plan to do the job ourselves. The gardener made a profit on the materials and received a healthy hourly return for his work.

The quickest way to have a finished look is to buy sufficient sod to cover your entire lawn area. But this is expensive. Our gardener bought enough sod to cover about 1/3 of the area and, by distributing it well, the entire lawn was covered within three months. The most reasonable way would have been to plant cuttings. But the time needed to complete the lawn would have increased by several months.

Specific Steps to Growing a Lawn

If you decide to grow a new lawn yourself, here are some specific steps suggested by Winona Guidry in *Enjoy Your Home Grounds*:

1. The first work to do is to level the outdoor area so as to have surface drainage away from the house. If you remove some of the topsoil, save and use it. Be careful when leveling under trees.
2. Plow, grade, and harrow. You may have to haul soil to fill in low spots.
3. Add organic matter, such as barnyard manure, peat, compost, well-decomposed sawdust, gin trash, or well-rotted rice hulls. For an average 75 × 150 feet, use 1 ton of barnyard manure or equivalent and 150 pounds of 0-12-12 or 0-14-7 fertilizer. If you do not add organic matter, use 150 pounds of 3-12-12 or 6-8-8. Work it well into the soil.
4. Rake until the seedbed is very fine, then roll.
5. You are now ready to seed, sprig, or sod your lawn.

Maintaining Your Lawn

Once you have a pretty lawn, how do you go about maintaining it? You can do the work yourself but you will need a lawn mower and edger. It may require cutting once a week in the growing season. A relatively small lawn is easy to maintain and is a splendid form of exercise. Amos Alonzo Stagg, the great University of Chicago football coach, was still cutting his grass at age 99.

Another solution is to have a gardener take care of your yard. The charge in our area for a weekly visit is $30 to $40 a month. The quality of work ranges from excellent to adequate.

My own view is that there is no substitute for personal attention. It can be argued that your time is valuable and could be better utilized elsewhere. This is true in many cases on a purely monetary basis, but doing your own landscaping can be looked on as a delightful change of pace and relaxation. My wife takes considerable interest in our yard and recently has been cutting the lawn for exercise. She also takes care of the shrubs and flowers. My job is to maintain the pool.

A third approach, perhaps in combination with one of the others, is to utilize a lawn care company. Such a firm treats your lawn a specified number of times a year. The yearly cost for doing this work is normally based on the square footage involved. For example, if a lawn covers 5,000 square feet, the charge may vary from $150 to $200. The company would usually make four to five visits during the growing season.

If you decide to use a lawn care company, be sure to obtain all the facts. Points you may wish to determine include: Is the company reliable? Will they do a good job? Do you have the right to terminate the agreement without penalty?

A Garden of Your Own

What about having your own fruit and vegetable garden? With the high cost of living, this can be a great way to save money. It also should be of interest to the health-minded and the environmentalists. This sounds great in theory but can be demanding, discouraging, and expensive in some areas. A neighbor in New Orleans tried a small vegetable garden, specializing in tomatoes. He found, however, that before he could apply protective measures insects had eaten the green fruit. After six months, he gave it up as an impossible task.

In contrast, a friend who lives nearby has successfully kept a garden for years and spends much of his spare time with this hobby. If you are interested, I would suggest you contact your local county agent for advice on maintaining your soil and to learn which crops do well in your locality.

A Garden for the Children

What about a garden for youngsters? Today it is a real challenge to direct their energy toward projects that can be useful in adult life. An appreciation of the soil is certainly a worthwhile undertaking. There is an interesting article in the March 1973 *South Central Bell Notes* pointing out how to start a small flower garden. This could apply to a vegetable or fruit garden.

> Looking for something different or unusual to do with your child for coming holidays and vacations? One of the most stimulating things you can do is help your child plan and plant his first garden. Observing how living things grow will be rewarding and exciting.
>
> A child's first garden should be a small one, about 3′ × 6′, with flowers that are easy to grow. First choices should be annuals. They must be planted each year, but the rewards of your youngster's first efforts are almost immediate.
>
> In choosing the flowers to plant, the marigold is a popular choice, with many varieties available. Other favorites are zinnias and petunias.

If your child likes flowers for their smell as well as their color, sweet alyssum would be a good choice.

All these plants do well even in poor soil and thrive on lots of sunlight. Most will bloom from seed in six to eight weeks and will last until first frost.

Even a small garden will require weekly care. After preparing the soil, planting, and marking the rows, watering and weeding will keep a youngster busy, but happy and proud as he is rewarded with flowers from his private domain.

Our Introduction to Gardening

Upon our arrival at the United States Military Academy, a neighbor knocked on our door one evening and said, "My name is Jack Meyer; welcome to West Point." Jack presented us with a bowl of fresh strawberries and a basket of tomatoes.

We spent a delightful evening and learned a great deal about what to expect during our three year tour at the Academy. Before leaving, he told us of a garden plot directly behind our backyard and encouraged us to plant some fruits and vegetables.

The next afternoon, he showed us how to go about it. There were twelve plots being worked by officers who had homes in the area. Jack had one of them and his crops were the most productive. Under his guidance, we started our first garden. Several months later we were eating tomatoes, strawberries, and muskmellon.

The following year, we moved into a house in another area at West Point. We took along some of our strawberry plants and newly found gardening knowledge. There was a sizable area for gardening in back of our new quarters. I rushed out to get started. The soil seemed different but I didn't think much about it—even though no one else was using it. It was rocky in contrast to the rich black earth near our previous house. Shortly after planting my strawberries, I came down with a terrible case of poison ivy. After hospitalization, I tried diligently to achieve results but did not succeed.

Roses and Other Flowers

Roses are particularly appealing to our family. We have two plants in our backyard but have had only mediocre results. Two horticulturists pointed out their appeal and the work involved.

> Roses are one of the most interesting and popular of all garden plant materials. For centuries, the universal appeal of the rose has been due to its versatile landscape uses as well as its cut flower quality. Few plants

> can supply color for such an extended period. . . . Do not consider growing roses unless you are willing to devote special attention to the control of insects and diseases. No factor in growing roses is more important than a systematic program for insect and disease prevention and control . . . both insects and disease pests can cause real concern. Yet, for individuals who are truly fond of roses, the pleasure of having roses will justify the expense and time required to keep insects and disease problems to a minimum.[1]

I have a good friend who has been very successful in raising roses and lectures on the topic. He indicated the best way to start is to purchase only two or three bushes. A bush can be bought for as little as 99 cents, but one of fine quality costs $5 to $6. Wes Strauch emphasized that to have a first class rose garden "you must devote a lot of time and study—you must be all wrapped up in it."

There are many other flowers available for those individuals who may wish to accept a lesser challenge than the rose. A good deal of research in recent years has made it easier to raise beautiful flowers. Frederick C. Klein writes that:

> Professional flower breeders are applying techniques previously used in the development of food crops to produce plants that are more vigorous, uniform, and disease-resistant and are freer-flowering and easier to grow than anything that has been available before. Roughly two-thirds of the flower varieties that home gardeners will plant this spring didn't even exist 20 years ago. . . .
>
> In years past, many American gardens consisted primarily of such perennial bloomers as roses, lillies, phlox, and delphiniums, whose success depends largely on the gardener's skill in periodically pruning and dividing his plants and shielding them against bad weather. This sort of gardening continues, of course; flower fanciers tend to gravitate to these more challenging types once they have mastered simpler ones.
>
> Increasingly, though, gardeners are planting annual flowers from seed or purchased plants, and they are getting results that once went only to the highly skilled. "It used to take a green thumb to have a good garden; now all it takes is a dirty hand," says William Carlson, associate professor of horticulture at Michigan State University and executive secretary of the National Association of Bedding Plant Growers.[2]

[1] Newil G. Odenwald and Claude Blackwell, *Roses*, Cooperative Extension Service, LSU, Baton Rouge, Louisiana, pp. 14-15. (Courtesy of Louisiana Cooperative)

[2] Excerpt from the *Wall Street Journal's* article of April 30, 1973

Figure 9-2a Suspense—a top quality rose that anyone can grow.

Figure 9-2b Blue Moon—a rosarian's dream, a blue rose.

Figure 9-2c Dainty Bess—a modern variety of an old fashioned rose.

Figure 9-2d Peace—rose of the century. (Courtesy of Weston G. Strauch)

This article also points out that although "no one knows exactly how many Americans tend a patch of flowers during the summer . . . estimates range as high as 20 million." An incentive to some is the W. Atlee Burpee Company offer of $10,000 to the person who finds a pure white marigold in her garden and sends in seed that can produce a similar one.

Landscaping Expenditures

What should you budget for landscaping? This will vary for each family. You may wish to note our initial expenses:

INITIAL COSTS

Grass	$ 57
Fertilizer	48
Soil (3 loads)	39
Gardener	125
Plants	108
Seeds	15
Edger (electric)	45
Lawnmower (electric)	100
Basic gardening tools (rake, shovel, broom, hoe, trowel, cultivator)	18
Hoses, nozzles, sprinkler	49
Total initial costs	$604

After the landscaping is complete, there is yearly maintenance. Expenditures will depend upon how much work you do yourself, the size of your yard, where you live, and climatic conditions.

ANNUAL UPKEEP

Plants	$ 40
Seeds	12
Fertilizer	20
Tools	15
Water	12
Repairs	20
Total annual cost	$119

You may wish to keep detailed records of your landscaping costs and a status report of work being done by the family. One method of listing this information is presented in Figure 9-3.

RECORD OF COST

(Include items bought and work paid for.)

DATE	WHAT I BOUGHT	EST. VALUE OF PLANTS I HAD OR WAS GIVEN	VALUE TO DATE
EXAMPLE: February 2	2 Abelias $1.20	5 Crape Myrtles $5.00	$6.20

RECORD OF FAMILY HELP ON PROJECT

DATE	WHO HELPED	WHAT THEY DID
EXAMPLE: April 10	Mother	Helped set chrysanthemum plants

Figure 9-3 Formats of records you may wish to maintain. (Guidry, Winona, *Enjoy Your Home Grounds*, Cooperative Extension Service, LSU, Baton Rouge, Louisiana, p. 10. Courtesy of Louisiana Cooperative.)

Summary

Landscaping your home should be considered at the planning stage. By using this approach, you can determine and set aside the money required for this purpose. It also permits you to schematically look at the total picture, including the house, patio, driveway, pool, and lawn. As a result of such a study, you should be able to decide what type of landscaping is most appropriate. At this point, you can make changes, at no cost, and obtain a home that satisfies your requirements from an artistic and practical view point.

In developing your landscaping plans, you should consider the current and future needs of your family as well as the possibility of one day selling your home. Determine the type grass, trees, shrubs, and flowers that are best suited to your locality. You should also give thought to a fruit and vegetable garden for you and perhaps the children.

Upkeep for your landscaping must also be considered. How expensive will it be to hire a gardener or lawn care firm? Can I do it myself? How much will the initial equipment cost?

Contact your county agent or your land grant university—either can be of great help at no expense to you. They can provide you with pamphlets on all aspects of landscaping.

CHAPTER 10
Maintenance, Repairs, Improvements

You should give consideration to the upkeep of your new home shortly after you are settled. Good maintenance will reduce your need for repairs. As the house grows older, expenditures for repairs will increase. This chapter will also discuss the subject of home improvements. Ideally, in building your home, you should make plans for later additions. For example, a neighbor built his home when he had need for only one bedroom. To conserve limited funds, he completed the first floor that included the master bedroom. He decided he would finish the second floor after the first child arrived.

Maintenance

Maintaining your house means to keep it in a certain condition or position of efficiency and good repair. If you have done a quality job in building your home, your maintenance should be minimal. In the initial planning, we pointed out the importance of having a home that will be easy to take care of.

As you look at your present abode, are you faced with sizable repair bills because of poor quality construction? Likewise, have you contributed to this expense by failing to keep your premises and equipment in good shape?

Preventive maintenance, taking proper care of your equipment prior to its breaking down, could preclude many major repairs. The advantages are two-fold; you are able to use the equipment longer and it reduces your repair bills. The key to it all is that you must initially buy good equipment and have it properly installed. The best maintenance will be wasted if the construction is poor and the equipment inferior. If this is true, your best solution is to replace it.

Maintenance can be increased considerably by poor planning. For example, at a nearby home completed last year, the owner wanted a patio that would be well protected. The architect devised a rectangular area surrounded on three sides by the house and on one side by the garage, completely enclosing the area except for the open ceiling. Upon seeing it finished, the owner was delighted. But this happiness was short-lived. After the first heavy rain, the patio was flooded. No drainage had been provided and the only way to remove the water was to bail it out. The lowest estimate to remedy the situation was $950. He decided to live with it for six months and, during the rainy season, had a serious maintenance problem.

From a maintenance standpoint, there is an advantage to building your own home. Your presence enables you to check on the quality and ask questions about keeping the equipment in good condition. Make a point to find out from the experts how best to take care of your roof, rugs, floors, air conditioning, and other items. Proper maintenance can extend their life considerably.

Tools and Storage

In order to maintain your equipment properly, you need adequate tools and a satisfactory place to store them. In planning your home, provide adequate space for this purpose. My recommendation is to use space within an enclosed garage, basement, or utility room. You may wish to purchase a small shed[1] that could be located in your backyard. We keep our tools in the dressing room of our garage, adjacent to the swimming pool. A pegboard makes them readily available; and shelving was built immediately below the board to store nails, tape, oil, and other supplies. You should have a nice store of nails and lumber to start with since all extra supplies belong to you.

Arrange your tools neatly and keep them in first class condition by cleaning and oiling them frequently. My chief problem is that my son and daughter are constantly using them to repair their bicycles and never return them to the proper place.

Tools are expensive and should be protected. If they're in a garage, keep it locked. Buy only the necessary items to do your maintenance and repair jobs satisfactorily. The cost of quality tools, including a box, for the routine tasks varies between $100 and $200. The items that I use most frequently are a variety of saws, hammers, screwdrivers, pliers, wrenches, and drills.

[1] A 12 x 18-foot lawn building costs $400 to $500 and can store considerable equipment. Extra items designed for this building include shelving, tool rack, pegboard, and work bench.

Record Keeping

The file system you developed in the planning stage of your home should include folders for maintenance, repairs, and home improvements. Keep warranties and instructions on how to care for your equipment. You may also wish to devise a checklist (see Figure 10-1) that could serve as a reminder to see that these items are properly maintained. You might include such equipment as the heating and cooling unit, floors, fence, refrigerator, oven, dishwasher, termite inspection, and painting.

Item to be Maintained	*Frequency of Maintenance*	*Date Last Accomplished*
Dishwasher	________	________
Flooring	________	________
Heating and cooling	________	________
Painting, exterior	________	________
Oven	________	________
Painting, interior	________	________
Refrigerator	________	________
Termite inspection	________	________

Figure 10-1 Checklist of items and frequency of maintenance.

Prompt Corrective Action

Maintenance should be prompt; and by utilizing quality materials, you simplify the upkeep. Hinges on your gates, for example, should be examined for loose screws. If gone unnoticed, a door could break off and repairs would be expensive. By doing simple jobs like oiling appropriate equipment and tightening the screws on your appliances and doors, you can keep them from becoming major problems. A wood fence should be examined periodically for loose knots. A little glue can remedy the situation. However, if the knot is lost, it means filling the hole with a substitute material. This will be more expensive and the replacement will not look as attractive.

Maintaining Your Cooling Unit

Central air conditioning is an expensive investment in warm areas—both in original cost and upkeep. In order to have it perform economically, it is important to keep the system operating efficiently. In Chapter 4, we emphasized the importance of adequate insulation. After you have a quality unit installed, what steps must be taken to maintain it? Our local utility advises its customers as follows:

> Clean or replace filter at least every month during the summer. Dirty filters put an added strain on your air conditioning, and on your cooling bills, too, so keep them clean.
>
> Set thermostat at highest setting that keeps you comfortable–and leave it there. The lower the setting, the longer your unit will run and the more it will cost to operate.
>
> Keep windows and doors closed–unnecessary opening of windows and doors lets in outside air that must be cooled and dehumidified.
>
> Keep hot sunlight out–use draperies, awnings, shutters, or shades to keep the sun out.
>
> Have system checked annually–a check of the mechanical operation of your system by a qualified serviceman is important for efficient, economical operation.[2]

Painting

You can save money if you paint your own home. It does, however, require considerable hard work and time. If you decide the interior needs painting, work in one room at a time over an extended period so that it does not become a major undertaking. In some families, the wife does the interior and the husband does the outside. In other cases, they work as a team.

Last year, I found it necessary to paint the outside of our house. We first checked on three firms that specialized in home painting and had a good reputation in the community. Their prices varied from $450 to $625. I then went to several job sites where painting was in progress; and after observing their work, I talked to several men about working for me in their spare time. Two came to my home and gave me estimates. In one case, I was told the job would cost $4.50 an hour if I supplied all the equipment, including brushes and paint. The second individual wanted to supply his own paint and equipment, his price was $6 an hour.

I next went to a first class paint company and priced brushes and paint. I also received good advice on the essential preparation prior to painting.

Painting the exterior of my house appeared to be a major project. Fortunately, my older son offered to paint if I would do the necessary preparation.

[2] "How the Weather Usually Means Higher Public Service Bills," *Homemaking*, New Orleans Public Service, June 1973, p. 3.

It sounded like a great idea, but this preparation was more than I bargained for. It took considerable time and effort.

You may be interested in the following price comparison that influenced my do-it-yourself decision.

Painting contractors	$450-$625
Painters	$288-$330
Our family	$63*

* Cost for brushes, sandpaper, paints, mildew remover, rollers, masking tape, etc.

Our experience proved that it is very important to buy quality paint, brushes, and materials. We bought the best grade paint at $7 per gallon. The price differential is minor and this extra cost is recouped in a longer lasting paint that is better looking.

Paint stores can match any color and they will shake the containers so they are ready for you to start painting. Make the best estimate as to the amount of paint you will use. If possible, buy a little less because you can always go back for more.

As indicated previously, my job was the preparation. This required removing the flaked paint. Fortunately, the paint on the house required very little sanding; but this was not true for our gates. I had made the mistake of applying two types of paint—latex and oil based. It was so loused up we had to burn it off after paint remover proved inadequate. In addition to removing the paint by burning and sandpapering, I was faced with caulking, puttying, securing loose nails, and removing mildew.

The final preparatory step prior to painting was to protect our brick, windows, doorknobs, shrubs, and driveway from paint splotches. Old sheets, newspapers, and masking tape did the trick. It is much easier to take these protective measures than to remove unwanted paint. Each evening after my son finished, I cleaned the brushes, removed paint from the ladder, and so on.

What about equipment? For outside painting, we found the oil-based paint gave us fine results. We bought quality brushes and used a roller wherever practical. The roller covered well and enabled my co-worker to complete the job quicker. Brushes were used on trim and any spots missed with the roller.

Painting requires good management. You want to make certain that the preparatory work, including purchasing the necessary equipment and cleaning up the area, has been done properly. You must also see that a first class job is accomplished.

Repairs

If you want to get an item repaired promptly and properly, you may find it pays to do it yourself. Today's abundant life style has brought a multitude of marvelous gadgets and equipment to many households. But how often have you heard:

"The stupid thing won't work!"

"Broken? I just bought it yesterday!"

"That serviceman promised to be here last week. What happened?"

In the future, American families will probably have even more material possessions and such problems can be expected to increase. In an interesting article on "The Horrors of Home Repair," *Life* magazine pointed up the difficulties. The author stated:

> Our craftsmen have become what amounts to a New Class: an elite body—inaccessible, accountable only to themselves and enjoying, for all their public indifference, an unprecedented growth in demand for their services. In doing so, they often seem to have done away with the once-proud ethics of their crafts: as manufacturers learned to apply obsolescence to their (products) craftsmen now appear to practice it, and generously, in their services. The object is to get in, get the money, and get out swiftly, leaving the pieces to fall behind them. The worse the job any one of them does, the sooner another one of them is called upon to redo it and the more work there is for everybody.[3]

A partial solution is the utilization of a very "available resource"—yourself. There used to be a belief in some quarters that white-collar employees shouldn't do blue-collar work. Such a philosophy meant that the college graduate didn't dirty his lily-white hands. Times have changed, with all economic levels attending universities; but there are still a goodly number in our society who make no effort to discover how useful their hands can be in making repairs.

It might be prudent to learn the basics of repairing electrical items, maintaining your car, handling a power mower without getting hurt, painting your home, or taking care of an outboard motor.

It amuses me to observe potbellies at the YMCA spending long periods in the steam room to work up a sweat and griping about "those lazy kids not showing up to cut the grass." Dr. White, the noted Boston heart surgeon, recommended mowing the lawn as a healthy form of exercise.

[3] William A. McWhirter, "The Horrors of Home Repair," *Life*, June 5, 1970, p. 58. (Copyright: *Life* magazine, © 1970-1972 Time Inc., William A. McWhirter)

If you expect to have the peace of mind of getting things done, it might be a good idea to learn more about the do-it-yourself philosophy. The benefits are not only the money saved and the physical exercise, but also the mental satisfaction of being able to do things yourself—and, in many cases, to do them better than the professional, because of personal interest and pride.

Replacements

From time to time, you will have to replace an item regardless of how well you maintain your equipment and how fine the quality. It will eventually wear out. Your own conscientious inspection system should enable you to approximate when a replacement is needed. This will permit you to more easily sell, trade, or replace the item while it is working.

In considering replacements, keep in mind that new and better products are constantly being made available to the public. Our nation is currently spending $30 billion annually on research and development in contrast to less than $1 billion 40 years ago. We can expect this research effort to continue apace. Keep abreast of the latest advances—not only for their convenience, but also in the event that you decide to sell your home.

Prior to making a change, check carefully into the newest equipment available. For example, the filter in our pool will soon have to be replaced. The present one uses diatomaceous earth and requires pulling up a plunger to backwash. This is a difficult and time-consuming chore. A new filter eliminates these problems. One manufacturer, Harmsco, states, " . . . we call it the Freedom Machine because that's what it gives you. Freedom from backwashing. Freedom from waste-water disposal. Freedom from four-way valve maintenance. Freedom from sand blowback. Freedom from diatomaceous earth. Freedom from separation tanks. And freedom from excessive water consumption." I am looking forward to buying a filter with these features.

Home Improvements

Let's assume that you want to make a major improvement in your home. If you have, for instance, initially built your home and completed only the first floor, you may now decide to finish the upstairs; or you may want to add an enclosed garage, den, or extra bedroom. These improvements can be best accomplished if they have been considered during the planning stage.

Once you decide to make a major improvement, you should follow the steps outlined in the first five chapters. The specific management information pertinent to building a swimming pool should also be reviewed. Take time to plan exactly what you want. In my capacity as a management consultant, a business executive recently asked, "Why take so much time to plan?" My response, "Although it requires considerable time, think back to all the hours you took to settle your last labor dispute. All the meetings, bitterness, and court action could have been avoided with adequate planning."

The time to allow for your home improvement project is during the planning stage. I was speaking with a successful contractor last week. He related the importance of making changes no later than during the blueprint stage:

> You can take an eraser to correct a rough drawing. Even if your design is finalized, you can modify it and a new set will cost only $50 to $75. But it really becomes expensive when the building is underway. Recently we contracted to add a new wing to an old house. The foundation had been poured. The framing, bricking, plumbing, and electrical wiring was also completed. At this point, the housewife decided to move her sink one foot to the right. She thought we were cheating her when our estimate indicated it would cost $1,200. I told her this change meant I had to contact the subcontractors involved for their changes. Numerous modifications would have to be made. If this was not done, the plumbing fixture would go right through the window. The electrician would have to modify his wiring. I needed to call back the carpenter to change the cupboards, etc. She finally realized her mistake in not taking greater time to plan her kitchen.

Adequate planning also means checking with care prior to selecting a firm to make a major improvement in your home. Try to obtain three estimates and secure a written contract.

People are usually given contracts at the most inopportune time—just before they are asked to sign them. They should be studied with care well in advance of closing the deal. If necessary, let your lawyer look it over.

A good friend of mine who failed to read his contract can tell you his sad experience:

> We decided to add a guest house in our backyard. I found the name of a contractor in the phone book and proceeded to call the owner of what I will call "Goodluck Industries." That evening he came to our house and quoted a price of $5,600. I signed the contract with the understanding that work would begin within 30 days. A week prior to his starting, I became concerned and proceeded to check him out with the BBB. To my dismay, I found that the owner of "Goodluck Industries" was a fraud. He never finished a job. But the contract specified that in the event of cancellation, I must pay a penalty of $560 even though no work was ever done. I went to my lawyer and after much

effort to include threatening legal action I bluffed my way out of it. But the time expended cost me $3,500 in lost business.

Mrs. Nell Weekly, head of the New Orleans Office of Consumer Affairs, offers seven excellent suggestions to prevent being taken in a home improvement swindle.

1. Get bids from several reliable contractors. Best sources of information on reliability are the Better Business Bureau (checking under both firm's name and contractor's name) and references from customers. Call the Remodelers Council and inquire about the arbitration provisions in their contracts. If this is important to you, make sure your contractor agrees to it;

2. Get it in writing, including details on work to be performed, guarantees, and specifications of materials to be used. Total cost and schedule of payments should also be in the contract. The most advantageous contract for the consumer reserves a substantial final payment to ensure that all work will be completed. Never make an advance payment for a special price or permit use of your home for advertising;

3. Do not sign contracts with door-to-door salesmen. Do not allow people to "inspect" your roof or heating system or your home for termite damage unless you have called them. Unnecessary work has been done and some homes actually damaged by roving "specialists";

4. Be wary of package deals which include financing. Shop as carefully for your money as you do for your contractor. Think twice before agreeing to mortgages; investigate FHA loans;

5. Make sure your contractor has a building permit;

6. Never sign a completion certificate until all work has been completed to your satisfaction;

7. For additional information, read: Chapter II, "How to Avoid Swindles on Home Improvements" in *Consumer Swindles and How to Avoid Them*; "Home Improvements for Love of Money" in the January issue of *Money Magazine*; and Better Business Bureau publications, including "Tips on Home Improvements" and Safeguard Bulletins 5 and 12.[4]

Budget Aspects

What do you budget for home improvements? Ideally you should save up enough money to pay the entire cost. Why pay a high rate of interest? If you cannot pay cash, let us look at the terms offered by a major firm.

Sears offers three credit plans that can be used to make home improvements.

[4] "How to Avoid Being 'Had'," *Dixie Roto Magazine, Times Picayune*, May 6, 1973, pp. 37, 48.

Revolving Charge Account. The finance charge is determined by applying a periodic rate of 1.5 percent per month (annual percentage rate of 18 percent) to the average daily balance. A loan over $500 must be paid within 10 months. If paid within 30 days of billing date, there is no interest cost.

Easy Payment Plan. The annual percentage rate of the finance charges is 20 percent. You have up to 36 months to pay, depending upon what you purchase.

Modernizing Credit Plan. The annual percentage rate of the finance charge is 14.75 percent. You have up to 5 years to pay. See Figure 10-2 for an example of monthly payments. Note that on $2,000, your interest for 5 years amounts to $792.

Sears credit personnel have informed me that generally speaking the company does not require any security on small loans—$1,500 to $2,000. The decision is based in large measure on a person's character and standing in the community. In contrast, a savings and loan association currently charges 9½ percent in our area but requires a mortgage. As pointed out in Chapter 6, a loan company is offering home improvement loans from 14.1 percent to 15.75 percent, but it expects the borrower to take out a life insurance policy. Banks in this area charge approximately the same as Sears.

Sears makes improvements in such areas as fencing, central cooling, heating, roofing, windows, doors, guttering, siding, ceiling tiling, awnings, carports, carpeting, built-in dishwashers, blown-in insulation, bathrooms, water heaters, and kitchens. Their guarantees read as follows:

> All installation labor provided through Sears shall be performed in a neat, workmanlike manner in accordance with generally accepted trade practices. Further, all installation shall adhere to all local laws, codes, regulations, and ordinances. Customer shall also be protected by insurance relating to Property Damage, Workman's Compensation, and Public Liability. . . .
>
> In addition to any guarantee which may be extended to you covering the product you have purchased, Sears, Roebuck and Co., guarantees the workmanship involved in the installation of the product as follows: Should any defect appear in such workmanship within one (1) year from date of installation, Sears will upon notice from you, cause such defects to be corrected at no additional cost.

If you use a firm like Sears, they provide the loan and make the home improvements. But if you shop around, you may be able to obtain a lower

Use This Convenient Table To Find Your Monthly Payment

These Modernizing Credit Plan terms are applicable only to residents of **Kentucky, Louisiana, Mississippi** and **Missouri.**

If Cash Price (Including any sales tax and shipping charge), less Down Payment If Any *amounts to*	We Shall Add for FINANCE CHARGE	Amount Payable Monthly is	*If Cash Price* (Including any sales tax and shipping charge), less Down Payment If Any *amounts to*	We Shall Add for FINANCE CHARGE	Amount Payable Monthly is
$100.00 to $110.00	$ 15.30	$ 5.00	$ 800.01 to $ 820.00	$320.00	$18.75
110.01 to 120.00	19.10	5.00	820.01 to 840.00	328.00	19.25
120.01 to 130.00	23.20	5.00	840.01 to 860.00	336.00	19.50
130.01 to 140.00	27.70	5.00	860.01 to 880.00	344.00	20.00
140.01 to 150.00	32.65	5.00	880.01 to 900.00	352.00	20.50
150.01 to 160.00	38.00	5.00	900.01 to 920.00	360.00	21.00
160.01 to 170.00	38.40	5.50	920.01 to 940.00	368.00	21.50
170.01 to 180.00	41.90	5.75	940.01 to 960.00	376.00	22.00
180.01 to 190.00	43.20	6.25	960.01 to 980.00	384.00	22.50
190.01 to 200.00	45.60	6.50	980.01 to 1,000.00	392.00	22.75
200.01 to 210.00	49.30	6.75	1,000.01 to 1,020.00	400.00	23.25
210.01 to 220.00	50.40	7.25	1,020.01 to 1,040.00	408.00	23.75
220.01 to 230.00	52.80	7.50	1,040.01 to 1,060.00	416.00	24.25
230.01 to 240.00	55.20	8.00	1,060.01 to 1,080.00	424.00	24.75
240.01 to 250.00	57.60	8.25	1,080.01 to 1,100.00	432.00	25.25
250.01 to 260.00	60.00	8.50	1,100.01 to 1,120.00	440.00	25.75
260.01 to 270.00	62.40	9.00	1,120.01 to 1,140.00	448.00	26.25
270.01 to 280.00	64.80	9.25	1,140.01 to 1,160.00	456.00	26.50
280.01 to 290.00	67.20	9.75	1,160.01 to 1,180.00	464.00	27.00
290.01 to 300.00	69.60	10.00	1,180.01 to 1,200.00	472.00	27.50
300.01 to 310.00	72.00	10.25	1,200.01 to 1,220.00	480.00	28.00
310.01 to 320.00	76.45	10.50	1,220.01 to 1,240.00	488.00	28.50
320.01 to 330.00	81.05	10.50	1,240.01 to 1,260.00	496.00	29.00
330.01 to 340.00	85.80	10.75	1,260.01 to 1,280.00	504.00	29.50
340.01 to 350.00	90.65	10.75	1,280.01 to 1,300.00	512.00	29.75
350.01 to 360.00	95.65	11.00	1,300.01 to 1,320.00	520.00	30.25
360.01 to 370.00	100.80	11.00	1,320.01 to 1,340.00	528.00	30.75
370.01 to 380.00	106.05	11.00	1,340.01 to 1,360.00	536.00	31.25
380.01 to 390.00	111.45	11.25	1,360.01 to 1,380.00	544.00	31.75
390.01 to 400.00	114.40	11.50	1,380.01 to 1,400.00	552.00	32.25
400.01 to 410.00	120.00	11.50	1,400.01 to 1,420.00	560.00	32.75
410.01 to 420.00	125.70	11.75	1,420.01 to 1,440.00	568.00	33.25
420.01 to 430.00	128.20	12.00	1,440.01 to 1,460.00	576.00	33.50
430.01 to 440.00	134.70	12.00	1,460.01 to 1,480.00	584.00	34.00
440.01 to 450.00	140.80	12.00	1,480.01 to 1,500.00	592.00	34.50
450.01 to 460.00	144.00	12.50	1,500.01 to 1,520.00	600.00	35.00
460.01 to 470.00	150.25	12.50	1,520.01 to 1,540.00	608.00	35.50
470.01 to 480.00	156.65	12.50	1,540.01 to 1,560.00	616.00	36.00
480.01 to 490.00	160.00	12.75	1,560.01 to 1,580.00	624.00	36.50
490.01 to 500.00	169.85	12.75	1,580.01 to 1,600.00	632.00	36.75
500.01 to 510.00	176.65	12.75	1,600.01 to 1,620.00	640.00	37.25
510.01 to 520.00	183.60	12.75	1,620.01 to 1,640.00	648.00	37.75
520.01 to 530.00	187.20	13.00	1,640.01 to 1,660.00	656.00	38.25
530.01 to 540.00	197.85	13.00	1,660.01 to 1,680.00	664.00	38.75
540.01 to 550.00	201.60	13.25	1,680.01 to 1,700.00	672.00	39.25
550.01 to 560.00	209.00	13.25	1,700.01 to 1,720.00	680.00	39.75
560.01 to 570.00	216.50	13.50	1,720.01 to 1,740.00	688.00	40.25
570.01 to 580.00	224.20	13.50	1,740.01 to 1,760.00	696.00	40.50
580.01 to 590.00	228.10	13.75	1,760.01 to 1,780.00	704.00	41.00
590.01 to 600.00	236.00	13.75	1,780.01 to 1,800.00	712.00	41.50
600.01 to 620.00	240.00	14.00	1,800.01 to 1,820.00	720.00	42.00
620.01 to 640.00	248.00	14.50	1,820.01 to 1,840.00	728.00	42.50
640.01 to 660.00	256.00	15.00	1,840.01 to 1,860.00	736.00	43.00
660.01 to 680.00	264.00	15.50	1,860.01 to 1,880.00	744.00	43.50
680.01 to 700.00	272.00	15.75	1,880.01 to 1,900.00	752.00	43.75
700.01 to 720.00	280.00	16.25	1,900.01 to 1,920.00	760.00	44.25
720.01 to 740.00	288.00	16.75	1,920.01 to 1,940.00	768.00	44.75
740.01 to 760.00	296.00	17.25	1,940.01 to 1,960.00	776.00	45.25
760.01 to 780.00	304.00	17.75	1,960.01 to 1,980.00	784.00	45.75
780.01 to 800.00	312.00	18.25	1,980.01 to 2,000.00	792.00	46.25

For purchases in excess of $2,000., add 40% of the Cash Price less any Down Payment for **FINANCE CHARGE** and divide by 60 for monthly payment.
The **ANNUAL PERCENTAGE RATE** of the **FINANCE CHARGE** will be **14.75%.**

Terms Card 50a Prepared by F. T. Weimer D/753 . . Sears, Roebuck and Co., 925 S. Homan Ave., Chicago, Ill. 60607 Sears

Figure 10-2 Sears modernizing credit plan terms. These terms apply only to specific home materials such as plumbing, heating, central air conditioning, kitchen cabinets, roofing, siding, etc. (Reprinted with permission of Sears, Roebuck and Co. © Sears, Roebuck and Co. 1973) Courtesy of Loan Guaranty Service Veterans Administration

priced loan and have the job done for less money. Be sure you are dealing with reliable companies.

Summary

By making periodic checks and taking appropriate corrective measures, you should be able to keep repair expenditures to a minimum.

When repair time arrives, determine if you can do the job yourself. Read the instructions with care. If help is needed, obtain a reliable firm.

New items for the home are being developed each year. Prior to making significant repairs on old equipment, examine new products that may do a better job and make your home more salable.

Effective home planning takes future improvement projects into consideration. In making these improvements, use the same management concepts you did in building your home. This will enable you to do a better job at a lower cost.

Epilogue

This management approach worked for me; it may not be your cup of tea. You may decide to use a contractor to build your home. But keep in mind you should provide him with a good deal of *free* supervision.

Another approach was recently suggested by a friend: "You busted your gut—for what? We looked at a home last Friday and moved in Monday. It was three years old, landscaped, ideal for our needs, and in perfect condition. All we had to do was settle in." My friend's solution may be a better answer for many—at least a lot simpler.

A word of caution: keep in mind that there are disastrous risks in building your own home. I recommend that you do not try to duplicate this experience if you have no managerial and financial background. Examine your own limitations before starting such a venture.

Let's assume that after weighing the alternatives of buying a home or hiring a contractor, you decide to build. Successfully managing the building of your own home will require a conscientious effort—a lot of hard work. But the satisfaction you gain from finishing this project makes it all worthwhile.

APPENDIX 1
Architectural Drawings for the Stillman House

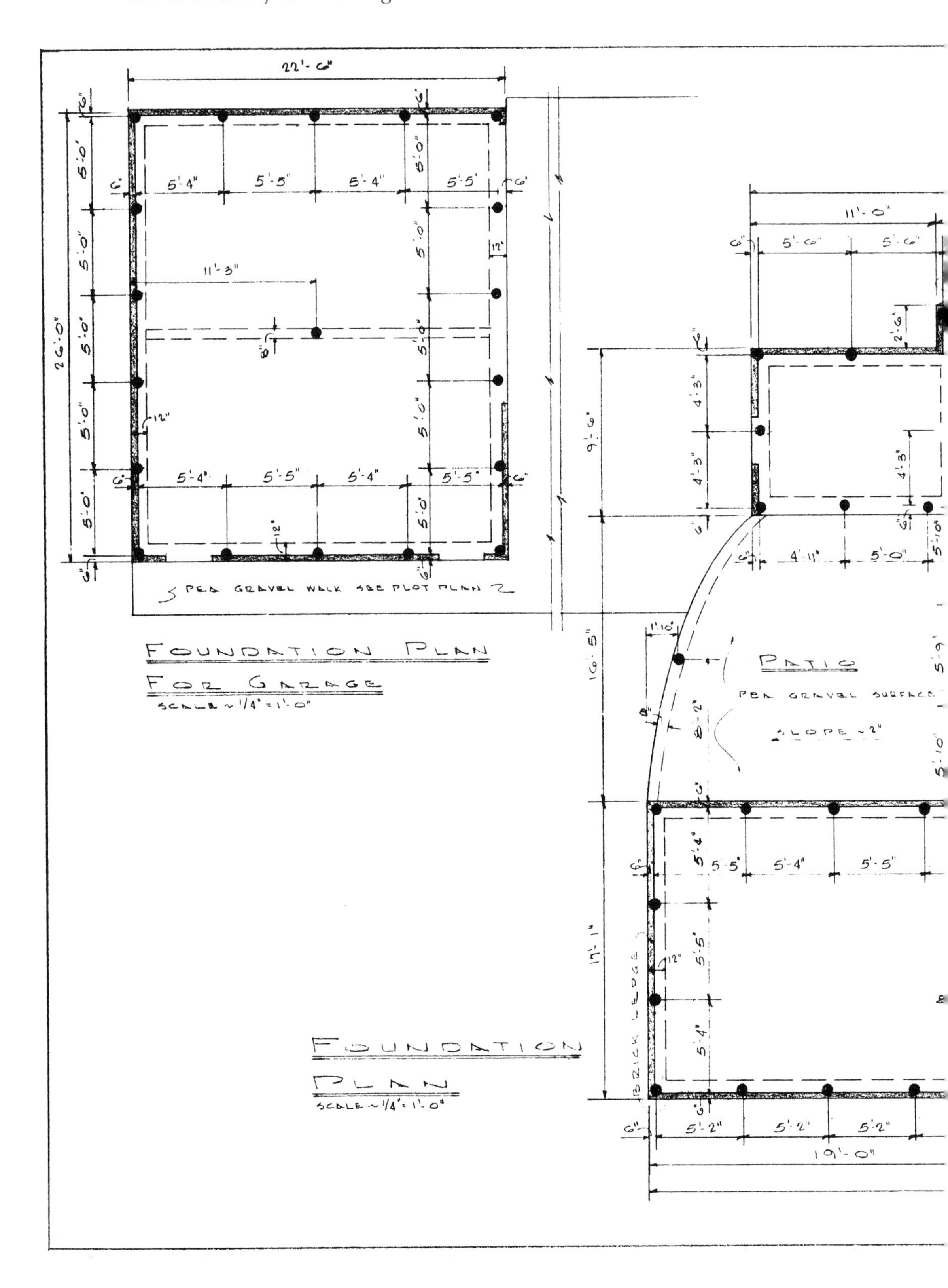

22'-6"
26'-0"
5'-4"
5'-5"
11'-3"
12"
PEA GRAVEL WALK SEE PLOT PLAN
FOUNDATION PLAN
FOR GARAGE
SCALE ~ 1/4" = 1'-0"
11'-0"
5'-6"
9'-6"
16'-5"
17'-1"
4'-3"
4'-11"
5'-0"
1'-10"
PATIO
PEA GRAVEL SURFACE
SLOPE ~ 2"
5'-4"
5'-5"
5'-2"
19'-0"
BRICK LEDGE
FOUNDATION
PLAN
SCALE ~ 1/4" = 1'-0"

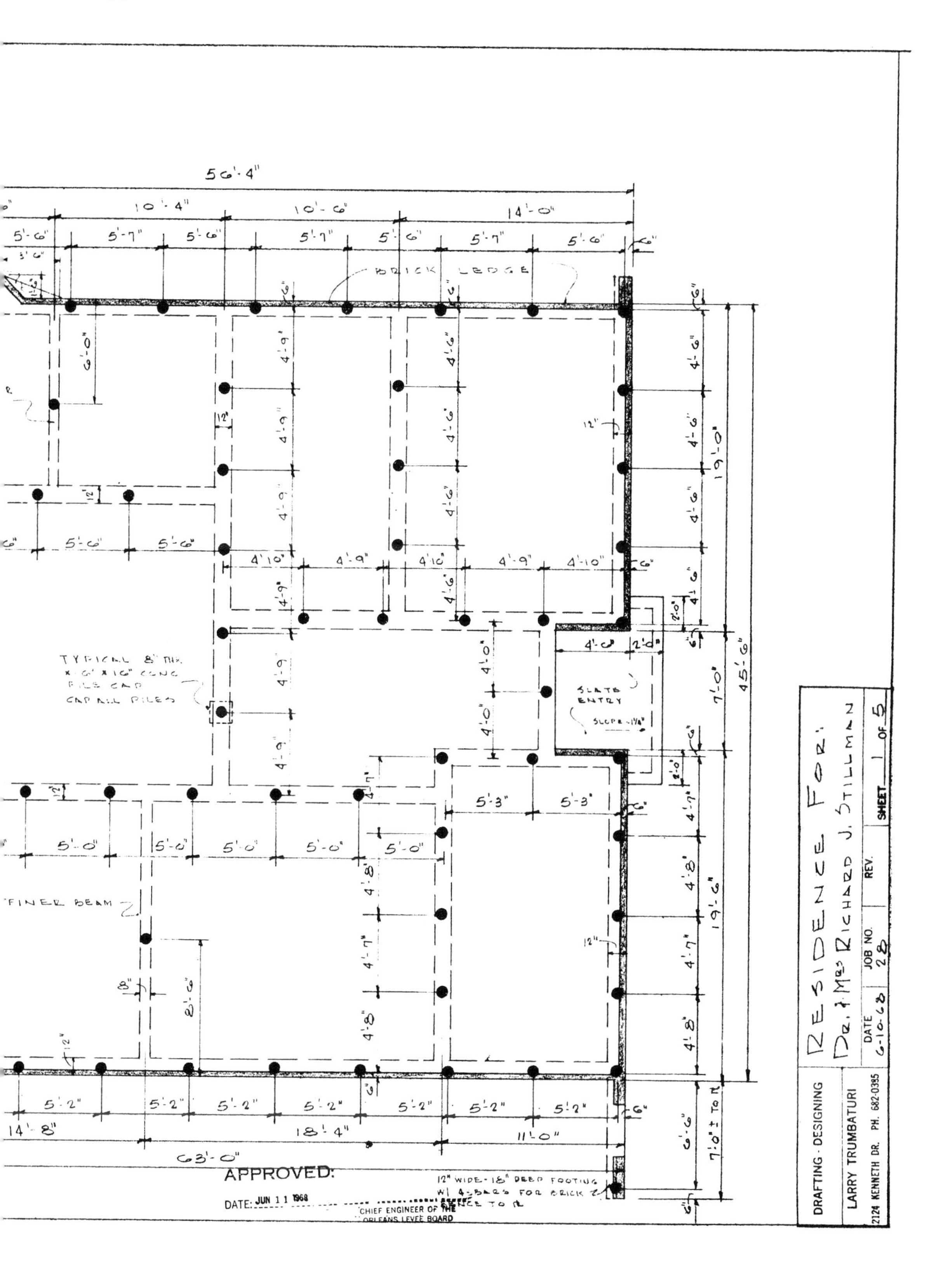
RESIDENCE FOR: DR. & MRS RICHARD J. STILLMAN
DRAFTING · DESIGNING
LARRY TRUMBATURI
2124 KENNETH DR. PH. 682-0395
DATE 6-10-68
JOB NO. 28
REV.
SHEET 1 OF 5
APPROVED:
DATE: JUN 11 1968
CHIEF ENGINEER OF THE ORLEANS LEVEE BOARD
BRICK LEDGE
SLATE ENTRY
TYPICAL 8" THK. x 6' x 16" CONC PILE CAP CAP ALL PILES
12" WIDE - 18" DEEP FOOTING W/ 4-BARS FOR BRICK FENCE TO R
56'-4"
63'-0"
45'-6"

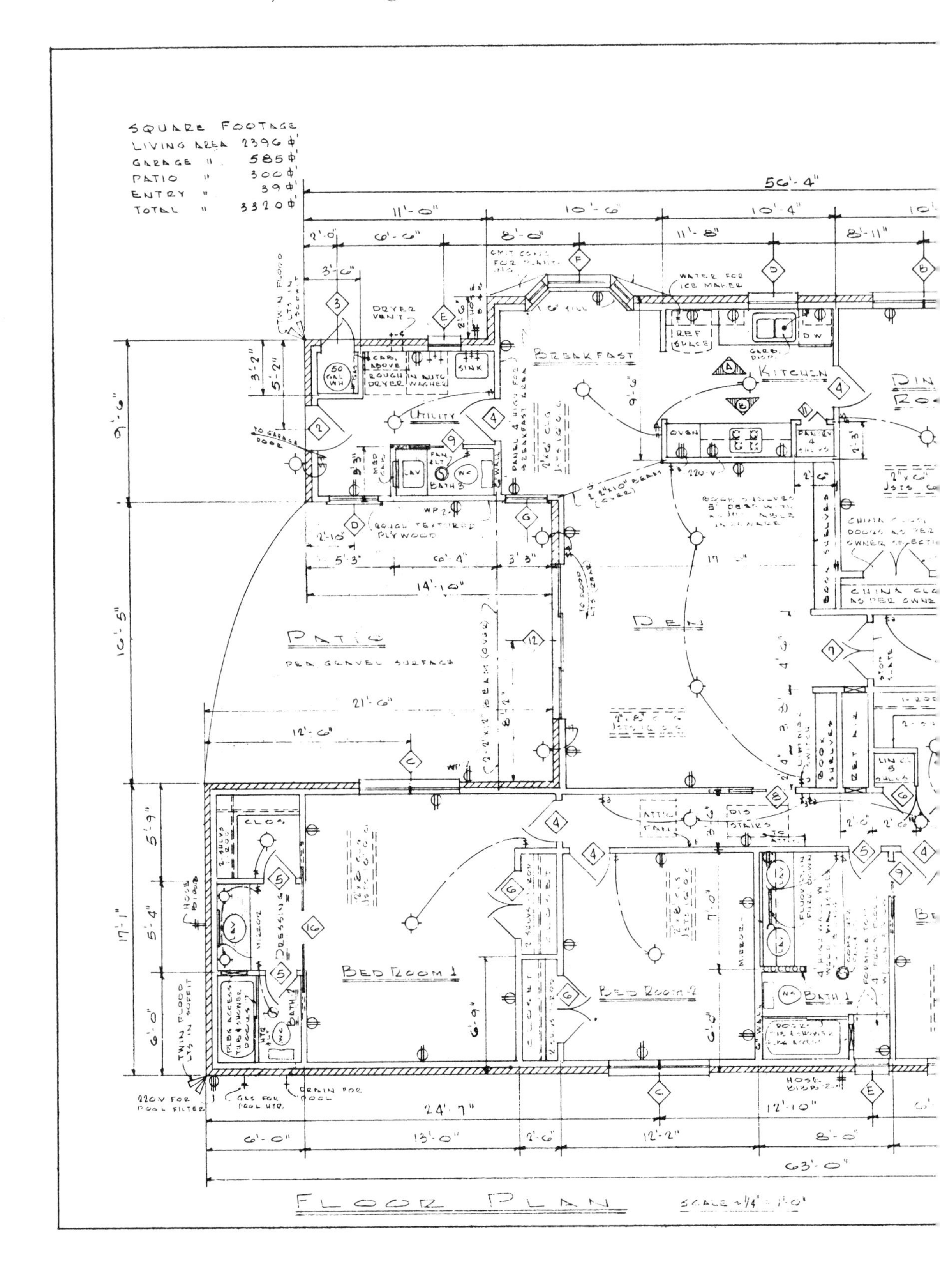
SQUARE FOOTAGE
LIVING AREA 2396 ☐'
GARAGE " 585 ☐'
PATIO " 300 ☐'
ENTRY " 39 ☐'
TOTAL " 3320 ☐'
56'-4"
11'-0"
10'-6"
10'-4"
DRYER VENT
WATER FOR ICE MAKER
BREAKFAST
KITCHEN
REF SPACE
DW
OVEN
PANTRY
UTILITY
SINK
50 GAL WH
BATH 3
LAV
WC
ROUGH TEXTURED PLYWOOD
14'-10"
PATIO
PEA GRAVEL SURFACE
21'-6"
12'-0"
DEN
ATTIC FAN
DIS STAIRS TO ATTIC
BOOK SHELVES
CHINA CLOSET
BEDROOM 1
DRESSING
CLOS.
BATH 2
BEDROOM 2
BATH 1
TWIN FLOOD LTS IN SOFFIT
HOSE BIBB
220 V FOR POOL FILTER
GAS FOR POOL HTR.
DRAIN FOR POOL
24'-7"
12'-10"
6'-0"
13'-0"
2'-6"
12'-2"
8'-0"
63'-0"
FLOOR PLAN
SCALE 1/4" = 1'-0"

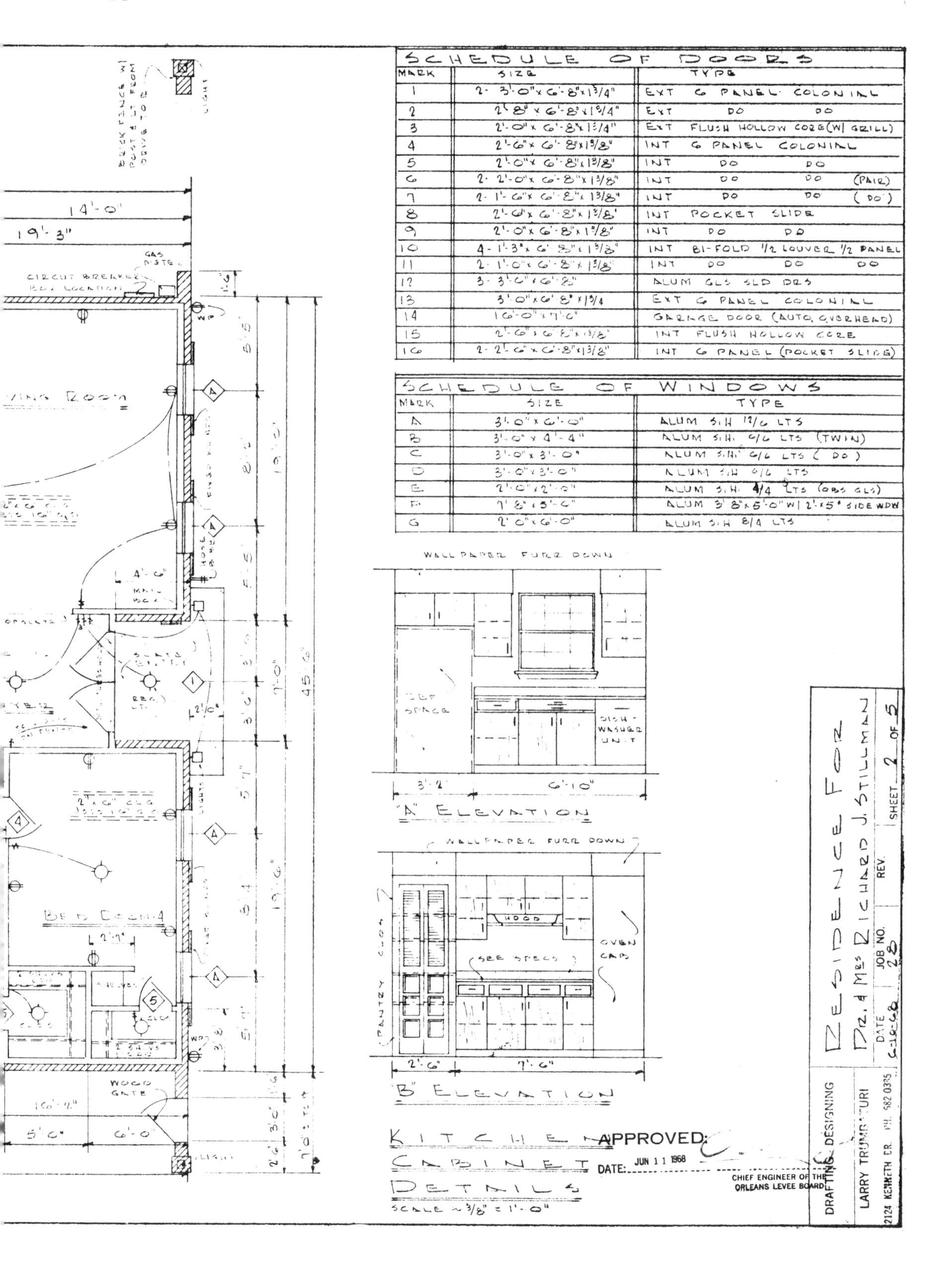

SCHEDULE OF DOORS

MARK	SIZE	TYPE
1	2- 3'-0" x 6'-8" x 1 3/4"	EXT 6 PANEL COLONIAL
2	2'-8" x 6'-8" x 1 3/4"	EXT DO DO
3	2'-0" x 6'-8" x 1 3/4"	EXT FLUSH HOLLOW CORE (W/ GRILL)
4	2'-6" x 6'-8" x 1 3/8"	INT 6 PANEL COLONIAL
5	2'-0" x 6'-8" x 1 3/8"	INT DO DO
6	2- 2'-0" x 6'-8" x 1 3/8"	INT DO DO (PAIR)
7	2- 1'-6" x 6'-8" x 1 3/8"	INT DO DO (DO)
8	2'-6" x 6'-8" x 1 3/8"	INT POCKET SLIDE
9	2'-0" x 6'-8" x 1 3/8"	INT DO DO
10	4- 1'-3" x 6'-8" x 1 3/8"	INT BI-FOLD 1/2 LOUVER 1/2 PANEL
11	2- 1'-0" x 6'-8" x 1 3/8"	INT DO DO DO
12	3- 3'-0" x 6'-8"	ALUM GLS SLD DRS
13	3'-0" x 6'-8" x 1 3/4	EXT 6 PANEL COLONIAL
14	16'-0" x 7'-0"	GARAGE DOOR (AUTO, OVERHEAD)
15	2'-6" x 6'-8" x 1 3/8"	INT FLUSH HOLLOW CORE
16	2- 2'-6" x 6'-8" x 1 3/8"	INT 6 PANEL (POCKET SLIDE)

SCHEDULE OF WINDOWS

MARK	SIZE	TYPE
A	3'-0" x 6'-0"	ALUM S.H 12/6 LTS
B	3'-0" x 4'-4"	ALUM S.H. 6/6 LTS (TWIN)
C	3'-0" x 3'-0"	ALUM S.H. 6/6 LTS (DO)
D	3'-0" x 3'-0"	ALUM S.H 6/6 LTS
E	2'-0" x 2'-0"	ALUM S.H. 4/4 LTS (OBS GLS)
F	7'-8" x 5'-0"	ALUM 3'-8" x 5'-0" W/ 2'-x 5' SIDE WDW
G	2'-0" x 6'-0"	ALUM S.H 8/4 LTS

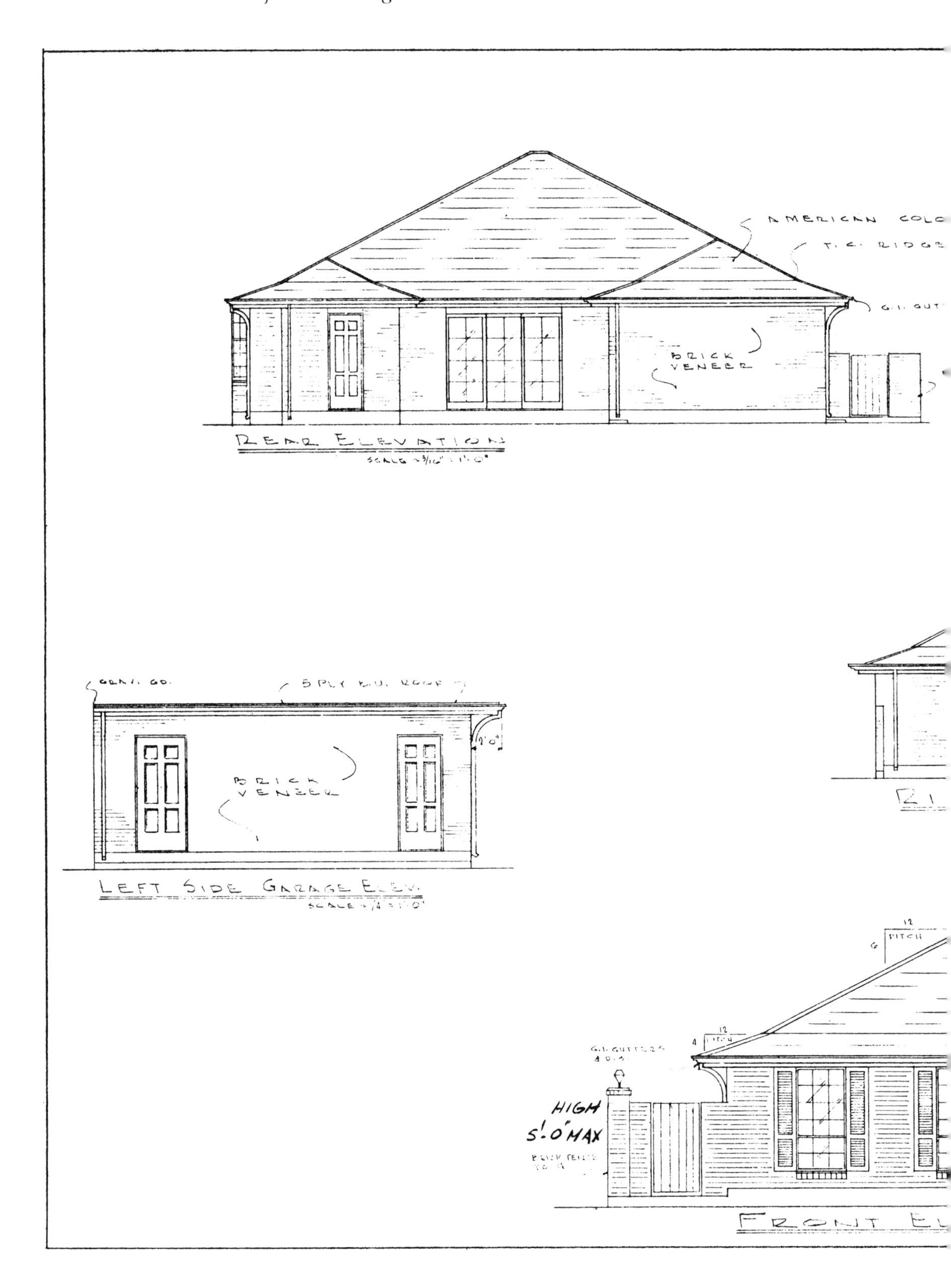
AMERICAN COLO
T. C. RIDGE
G. I. GUT
BRICK
VENEER
REAR ELEVATION
5 PLY B.U. ROOF
BRICK
VENEER
LEFT SIDE GARAGE ELEV.
12
6 PITCH
12
4 PITCH
G.I. GUTTERS
& D.S.
HIGH
5'-0" MAX
FRONT EL

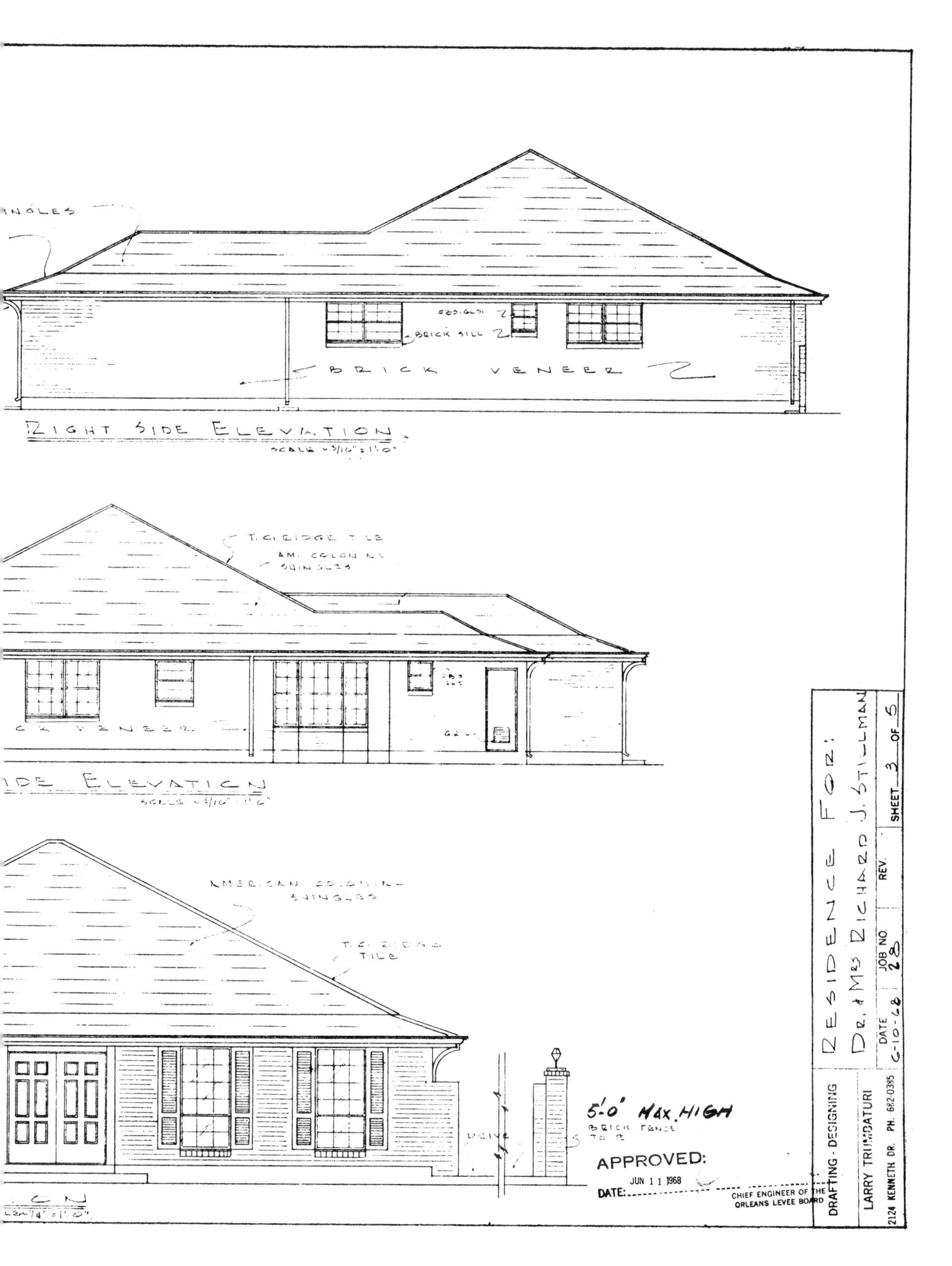
BRICK SILL
BRICK VENEER
RIGHT SIDE ELEVATION
T.C. RIDGE TILE
SHINGLES
TILE
5'-0" MAX. HIGH
APPROVED:
DATE: JUN 11 1968
CHIEF ENGINEER OF THE
ORLEANS LEVEE BOARD
RESIDENCE FOR:
DR. & MRS RICHARD J. STILLMAN
DATE 6-10-68
JOB NO. 28
REV.
SHEET 3 OF 5
DRAFTING - DESIGNING
2124 KENNETH DR. PH. 682-0385

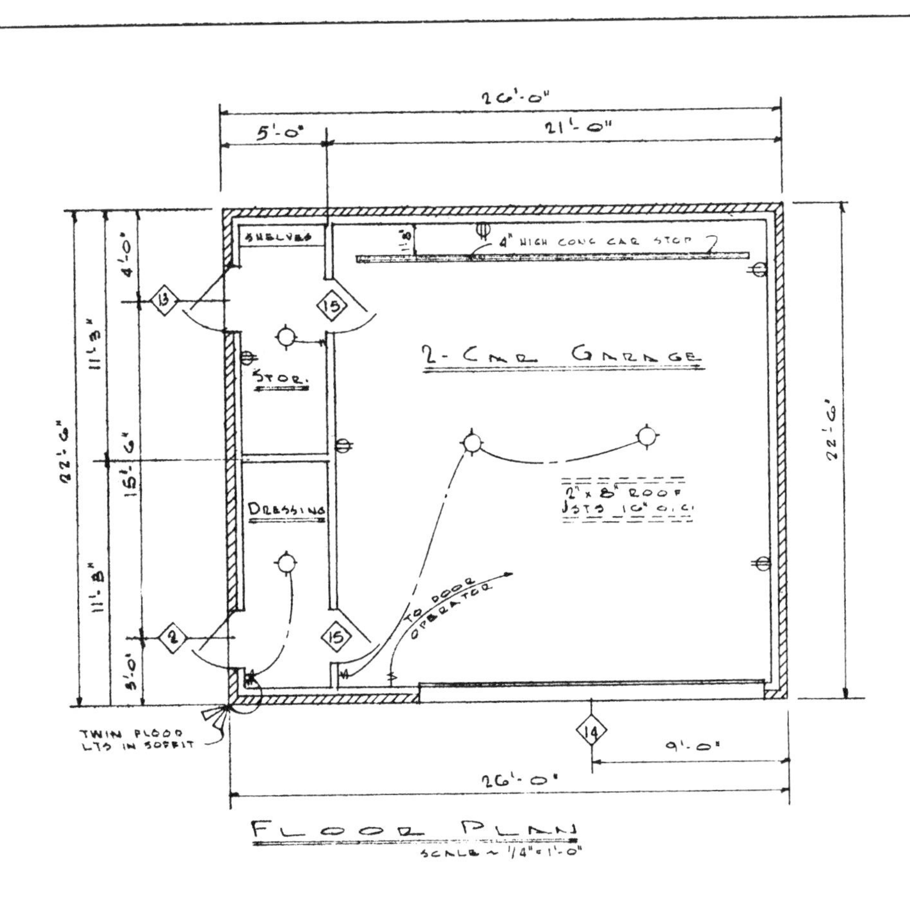
26'-0"
5'-0"
21'-0"
SHELVES
4" HIGH CONC CAR STOP
2-CAR GARAGE
STOR.
DRESSING
2" x 8" ROOF JSTS 16" O.C.
TO DOOR OPERATOR
TWIN FLOOD LTS IN SOFFIT
9'-0"
26'-0"
22'-6"
11'-3"
15'-6"
11'-8"
FLOOR PLAN
SCALE ~ 1/4" = 1'-0"

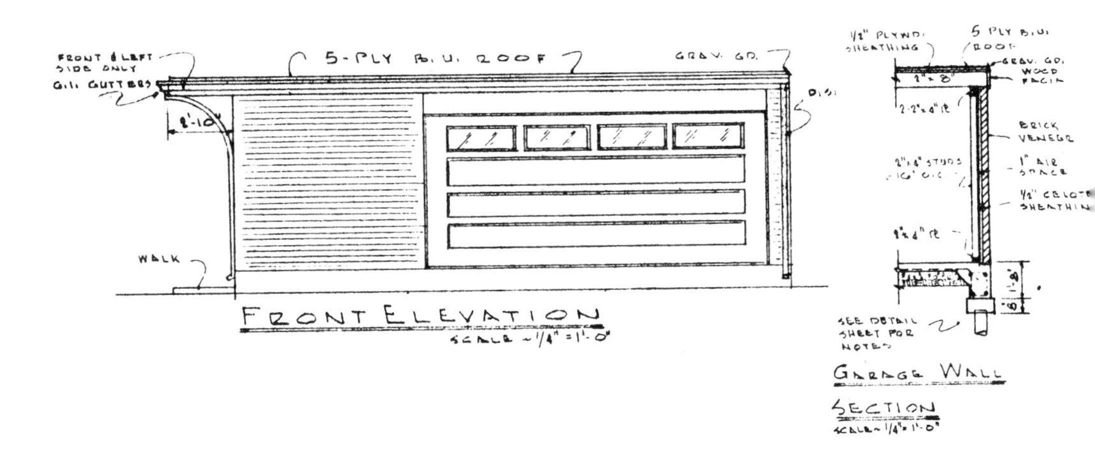
FRONT & LEFT SIDE ONLY
GAL GUTTERS
5-PLY B.U. ROOF
GRAV. GD.
WALK
FRONT ELEVATION
SCALE ~ 1/4" = 1'-0"
1/2" PLYWD. SHEATHING
5 PLY B.U. ROOF
GRAV. GD. WOOD FACIA
BRICK VENEER
1" AIR SPACE
1/2" CELOTEX SHEATHING
SEE DETAIL SHEET FOR NOTES
GARAGE WALL SECTION
SCALE ~ 1/4" = 1'-0"

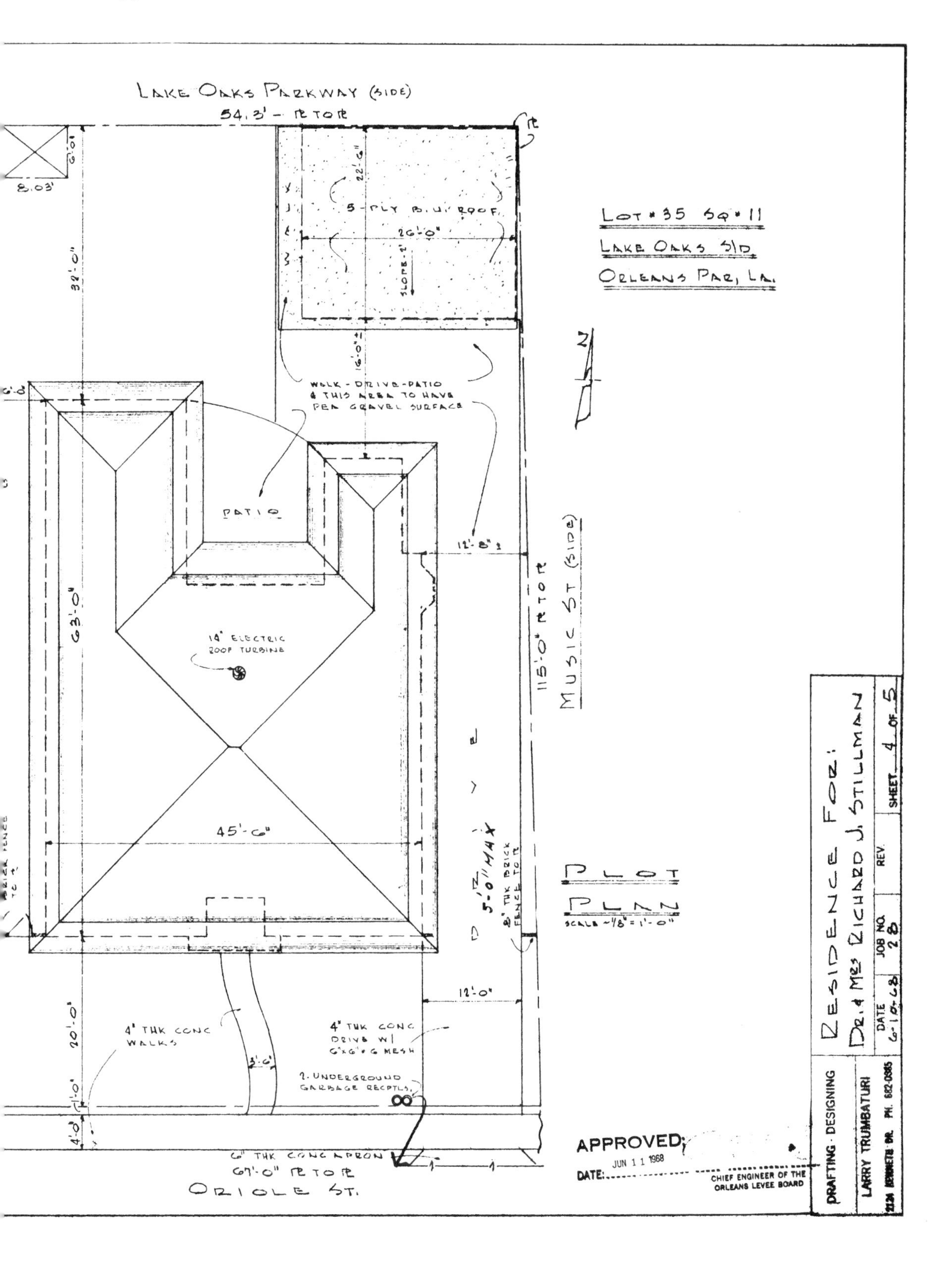
LAKE OAKS PARKWAY (SIDE)
54.3' - R TO R
LOT # 35 SQ # 11
LAKE OAKS S/D
ORLEANS PAR, LA.
5-PLY B.U. ROOF
26'-0"
22'-0"
32'-0"
WALK - DRIVE - PATIO & THIS AREA TO HAVE PEA GRAVEL SURFACE
PATIO
12'-8"
115'-0" R TO R
MUSIC ST (SIDE)
14" ELECTRIC ROOF TURBINE
63'-0"
45'-6"
5'-0" MAX
8" THK BRICK FENCE TO R
PLOT PLAN
SCALE - 1/8" = 1'-0"
12'-0"
20'-0"
4" THK CONC WALKS
4" THK CONC DRIVE W/ 6"x6" # 6 MESH
3'-6"
2. UNDERGROUND GARBAGE RECPTLS.
6" THK CONC APRON
67'-0" R TO R
ORIOLE ST.
APPROVED:
DATE: JUN 11 1968
CHIEF ENGINEER OF THE ORLEANS LEVEE BOARD
RESIDENCE FOR:
DR. & MRS RICHARD J. STILLMAN
DRAFTING · DESIGNING
LARRY TRUMBATURI
DATE 6-10-68
JOB NO. 28
REV.
SHEET 4 OF 5

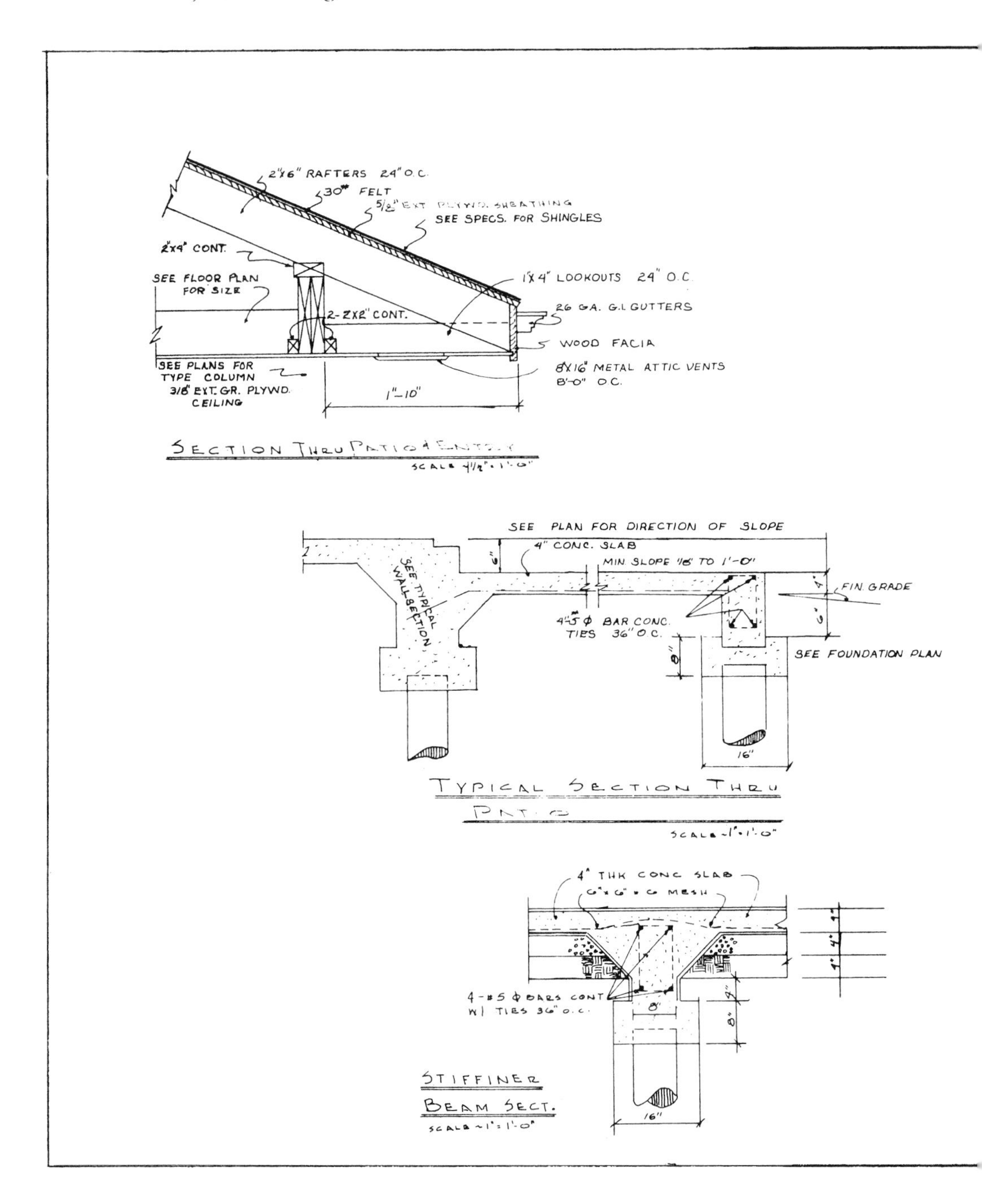
2"x6" RAFTERS 24" O.C.
30# FELT
SEE SPECS. FOR SHINGLES
2"x4" CONT.
SEE FLOOR PLAN FOR SIZE
2-2X2" CONT.
1"X4" LOOKOUTS 24" O.C.
26 GA. G.I. GUTTERS
WOOD FACIA
8"X16" METAL ATTIC VENTS 8'-0" O.C.
SEE PLANS FOR TYPE COLUMN
3/8" EXT. GR. PLYWD. CEILING
1"-10"
SECTION THRU PATIO & ENTRY
SEE PLAN FOR DIRECTION OF SLOPE
4" CONC. SLAB
MIN. SLOPE 1/8" TO 1'-0"
FIN. GRADE
SEE TYPICAL WALL SECTION
4-#5 Φ BAR CONC. TIES 36" O.C.
SEE FOUNDATION PLAN
16"
TYPICAL SECTION THRU PATIO
16"
STIFFINER BEAM SECT.

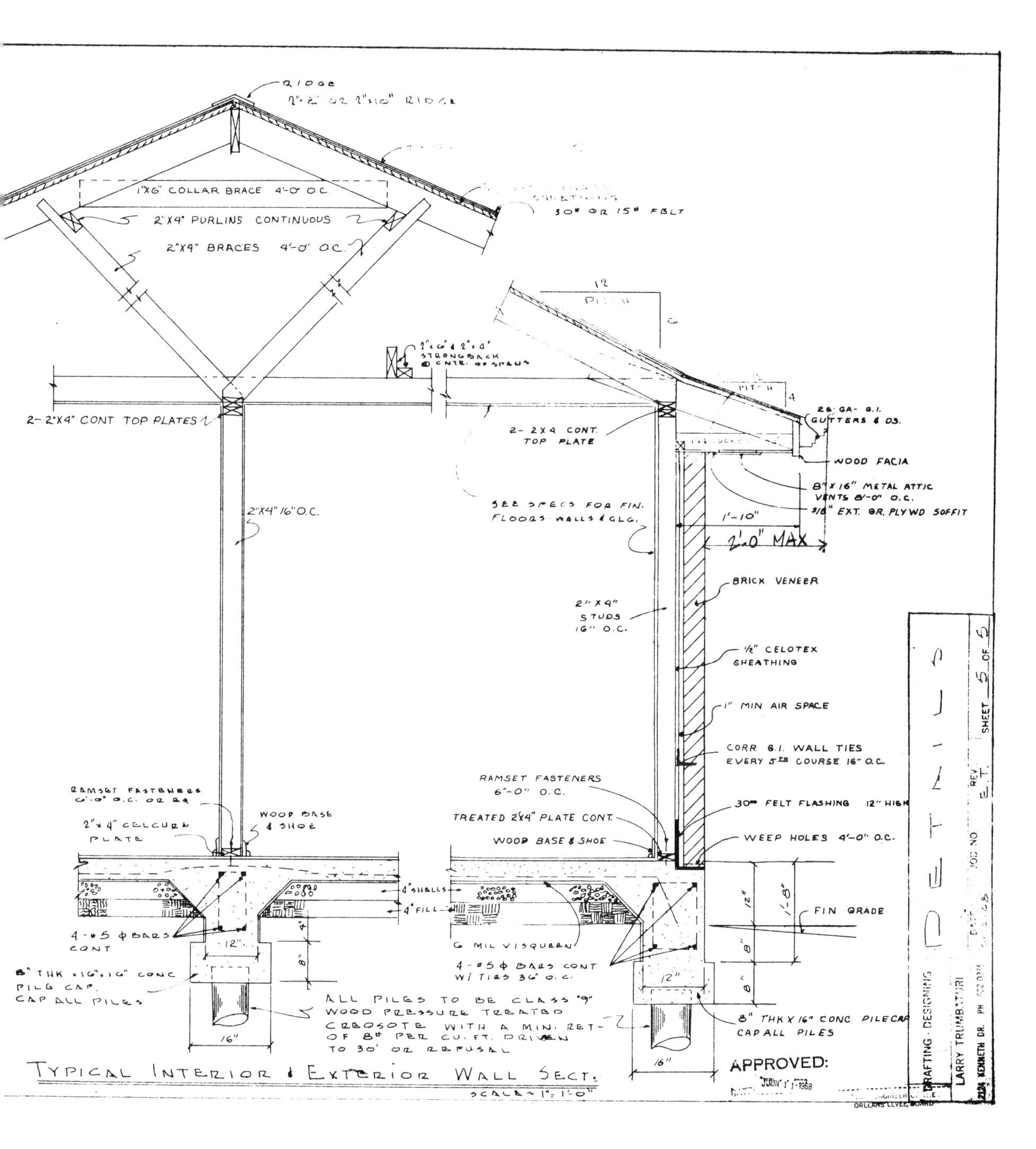
RIDGE
2"x8" OR 2"x10" RIDGE
1"X6" COLLAR BRACE 4'-0" O.C.
2"X4" PURLINS CONTINUOUS
2"X4" BRACES 4'-0" O.C.
30# OR 15# FELT
12
6
STRONGBACK
2-2"X4" CONT TOP PLATES
2-2X4 CONT. TOP PLATE
26 GA- G.I. GUTTERS & DS.
WOOD FACIA
8"X16" METAL ATTIC VENTS 8'-0" O.C.
3/8" EXT. GR. PLYWD SOFFIT
SEE SPECS FOR FIN. FLOORS WALLS & CLG.
2"X4" 16" O.C.
1'-10"
2'-0" MAX
BRICK VENEER
2"X4" STUDS 16" O.C.
1/2" CELOTEX SHEATHING
1" MIN AIR SPACE
CORR G.I. WALL TIES EVERY 5TH COURSE 16" O.C.
RAMSET FASTENERS 6'-0" O.C.
TREATED 2X4" PLATE CONT.
WOOD BASE & SHOE
30# FELT FLASHING 12" HIGH
WEEP HOLES 4'-0" O.C.
2"x4" CELCURE PLATE
4" SHELLS
4" FILL
FIN GRADE
4-#5 φ BARS CONT
6 MIL VISQUEEN
4-#5 φ BARS CONT W/ TIES 36" O.C.
8" THK x16"x16" CONC PILE CAP. CAP ALL PILES
ALL PILES TO BE CLASS "9" WOOD PRESSURE TREATED CREOSOTE WITH A MIN. RET. OF 8# PER CU. FT. DRIVEN TO 30' OR REFUSAL
8" THK X 16" CONC PILECAP CAP ALL PILES
12"
8"
16"
1'-8"
TYPICAL INTERIOR & EXTERIOR WALL SECT.
SCALE 1"=1'-0"
APPROVED:
JUN 1 1968
ORLEANS LEVEE BOARD
DETAILS
SHEET 5 OF 5
DRAFTING · DESIGNING
LARRY TRUMBATURI
KENNETH DR.

APPENDIX 2
Sample Description of Materials

U. S. DEPARTMENT OF HOUSING AND URBAN DEVELOPMENT
FEDERAL HOUSING ADMINISTRATION

For accurate register of carbon copies, form may be separated along above fold. Staple completed sheets together in original order.

☐ Proposed Construction

☐ Under Construction

DESCRIPTION OF MATERIALS

No. ______ (To be inserted by FHA or VA)

Property address ______ *City* ______ *State* ______

Mortgagor or Sponsor ______ (Name) ______ (Address)

Contractor or Builder ______ (Name) ______ (Address)

INSTRUCTIONS

1. For additional information on how this form is to be submitted, number of copies, etc., see the instructions applicable to the FHA Application for Mortgage Insurance or VA Request for Determination of Reasonable Value, as the case may be.
2. Describe all materials and equipment to be used, whether or not shown on the drawings, by marking an X in each appropriate check-box and entering the information called for in each space. If space is inadequate, enter "See misc." and describe under item 27 or on an attached sheet. THE USE OF PAINT CONTAINING MORE THAN ONE PERCENT LEAD BY WEIGHT IS PROHIBITED.
3. Work not specifically described or shown will not be considered unless required, then the minimum acceptable will be assumed. Work exceeding minimum requirements cannot be considered unless specifically described.
4. Include no alternates, "or equal" phrases, or contradictory items. (Consideration of a request for acceptance of substitute materials or equipment is not thereby precluded.)
5. Include signatures required at the end of this form.
6. The construction shall be completed in compliance with the related drawings and specifications, as amended during processing. The specifications include this Description of Materials and the applicable Minimum Property Standards.

1. EXCAVATION:

Bearing soil, type ______

2. FOUNDATIONS:

Footings: concrete mix ______; strength psi ______ Reinforcing ______

Foundation wall: material ______ Reinforcing ______

Interior foundation wall: material ______ Party foundation wall ______

Columns: material and sizes ______ Piers: material and reinforcing ______

Girders: material and sizes ______ Sills: material ______

Basement entrance areaway ______ Window areaways ______

Waterproofing ______ Footing drains ______

Termite protection ______

Basementless space: ground cover ______; insulation ______; foundation vents ______

Special foundations ______

Additional information: ______

3. CHIMNEYS:

Material ______ Prefabricated *(make and size)* ______

Flue lining: material ______ Heater flue size ______ Fireplace flue size ______

Vents *(material and size)*: gas or oil heater ______; water heater ______

Additional information: ______

4. FIREPLACES:

Type: ☐ solid fuel; ☐ gas-burning; ☐ circulator *(make and size)* ______ Ash dump and clean-out ______

Fireplace: facing ______; lining ______; hearth ______; mantel ______

Additional information: ______

5. EXTERIOR WALLS:

Wood frame: wood grade, and species ______ ☐ Corner bracing. Building paper or felt ______

Sheathing ______; thickness ______; width ______; ☐ solid; ☐ spaced ______" o. c.; ☐ diagonal; ______

Siding ______; grade ______; type ______; size ______; exposure ______"; fastening ______

Shingles ______; grade ______; type ______; size ______; exposure ______"; fastening ______

Stucco ______; thickness ______"; Lath ______; weight ______ lb.

Masonry veneer ______ Sills ______ Lintels ______ Base flashing ______

Masonry: ☐ solid ☐ faced ☐ stuccoed; total wall thickness ______"; facing thickness ______"; facing material ______

Backup material ______; thickness ______"; bonding ______

Door sills ______ Window sills ______ Lintels ______ Base flashing ______

Interior surfaces: dampproofing, ______ coats of ______; furring ______

Additional information: ______

Exterior painting: material ______; number of coats ______

Gable wall construction: ☐ same as main walls; ☐ other construction ______

6. FLOOR FRAMING:

Joists: wood, grade, and species ______; other ______; bridging ______; anchors ______

Concrete slab: ☐ basement floor; ☐ first floor; ☐ ground supported; ☐ self-supporting; mix ______; thickness ______";

reinforcing ______; insulation ______; membrane ______

Fill under slab: material ______; thickness ______". Additional information: ______

7. SUBFLOORING: ***(Describe underflooring for special floors under item 21.)***

Material: grade and species ______; size ______; type ______

Laid: ☐ first floor; ☐ second floor; ☐ attic ______ sq. ft.; ☐ diagonal; ☐ right angles. Additional information: ______

8. FINISH FLOORING: ***(Wood only. Describe other finish flooring under item 21.)***

Location	Rooms	Grade	Species	Thickness	Width	Bldg. Paper	Finish
First floor							
Second floor							
Attic floor ______ sq. ft.							

Additional information: ______

DESCRIPTION OF MATERIALS

9. PARTITION FRAMING:
Studs: wood, grade, and species ______ size and spacing ______ Other ______
Additional information: ______

10. CEILING FRAMING:
Joists: wood, grade, and species ______ Other ______ Bridging ______
Additional information: ______

11. ROOF FRAMING:
Rafters: wood, grade, and species ______ Roof trusses (see detail): grade and species ______
Additional information: ______

12. ROOFING:
Sheathing: wood, grade, and species ______ ; ☐ solid; ☐ spaced ______" o.c.
Roofing ______ ; grade ______ ; size ______ ; type ______
Underlay ______ ; weight or thickness ______ ; size ______ ; fastening ______
Built-up roofing ______ ; number of plies ______ ; surfacing material ______
Flashing: material ______ ; gage or weight ______ ; ☐ gravel stops; ☐ snow guards
Additional information: ______

13. GUTTERS AND DOWNSPOUTS:
Gutters: material ______ ; gage or weight ______ ; size ______ ; shape ______
Downspouts: material ______ ; gage or weight ______ ; size ______ ; shape ______ ; number ______
Downspouts connected to: ☐ Storm sewer; ☐ sanitary sewer; ☐ dry-well. ☐ Splash blocks: material and size ______
Additional information: ______

14. LATH AND PLASTER
Lath ☐ walls, ☐ ceilings: material ______ ; weight or thickness ______ Plaster: coats ______ ; finish ______
Dry-wall ☐ walls, ☐ ceilings: material ______ ; thickness ______ ; finish ______ ;
Joint treatment ______

15. DECORATING: ***(Paint, wallpaper, etc.)***

Rooms	Wall Finish Material and Application	Ceiling Finish Material and Application
Kitchen		
Bath		
Other		

Additional information: ______

16. INTERIOR DOORS AND TRIM:
Doors: type ______ ; material ______ ; thickness ______
Door trim: type ______ ; material ______ Base: type ______ ; material ______ ; size ______
Finish: doors ______ ; trim ______
Other trim *(item, type and location)* ______
Additional information: ______

17. WINDOWS:
Windows: type ______ ; make ______ ; material ______ ; sash thickness ______
Glass: grade ______ ; ☐ sash weights; ☐ balances, type ______ ; head flashing ______
Trim: type ______ ; material ______ Paint ______ ; number coats ______
Weatherstripping: type ______ ; material ______ Storm sash, number ______
Screens: ☐ full; ☐ half; type ______ ; number ______ ; screen cloth material ______
Basement windows: type ______ ; material ______ ; screens, number ______ ; Storm sash, number ______
Special windows ______
Additional information: ______

18. ENTRANCES AND EXTERIOR DETAIL:
Main entrance door: material ______ ; width ______ ; thickness ______". Frame: material ______ , thickness ______"
Other entrance doors: material ______ ; width ______ ; thickness ______". Frame: material ______ ; thickness ______"
Head flashing ______ Weatherstripping: type ______ ; saddles ______
Screen doors: thickness ______"; number ______ ; screen cloth material ______ Storm doors: thickness ______"; number ______
Combination storm and screen doors: thickness ______"; number ______ ; screen cloth material ______
Shutters: ☐ hinged; ☐ fixed. Railings ______ , Attic louvers ______
Exterior millwork: grade and species ______ Paint ______ ; number coats ______
Additional information: ______

19. CABINETS AND INTERIOR DETAIL:
Kitchen cabinets, wall units: material ______ ; lineal feet of shelves ______ ; shelf width ______
Base units: material ______ ; counter top ______ ; edging ______
Back and end splash ______ Finish of cabinets ______ ; number coats ______
Medicine cabinets: make ______ ; model ______
Other cabinets and built-in furniture ______
Additional information: ______

20. STAIRS:

Stair	Treads		Risers		Strings		Handrail		Balusters	
	Material	Thickness	Material	Thickness	Material	Size	Material	Size	Material	Size
Basement										
Main										
Attic										

Disappearing: make and model number ______
Additional information: ______

21. SPECIAL FLOORS AND WAIN JT:

	Location	Material, Color, Border, Sizes, Gage, Etc.	Threshold Material	Wall Base Material	Underfloor Material
Floors	Kitchen				
	Bath				

	Location	Material, Color, Border, Cap. Sizes, Gage, Etc.	Height	Height Over Tub	Height in Showers (From Floor)
Wainscot	Bath				

Bathroom accessories: ☐ Recessed; material ______; number ______; ☐ Attached; material ______; number ______
Additional information: ______

22. PLUMBING:

Fixture	Number	Location	Make	Mfr's Fixture Identification No.	Size	Color
Sink						
Lavatory						
Water closet						
Bathtub						
Shower over tub△						
Stall shower△						
Laundry trays						

△☐ Curtain rod △☐ Door ☐ Shower pan: material ______
Water supply: ☐ public; ☐ community system; ☐ individual (private) system.★
Sewage disposal: ☐ public; ☐ community system; ☐ individual (private) system.★
★*Show and describe individual system in complete detail in separate drawings and specifications according to requirements.*
House drain (inside): ☐ cast iron; ☐ tile; ☐ other ______ House sewer (outside): ☐ cast iron; ☐ tile; ☐ other ______
Water piping: ☐ galvanized steel; ☐ copper tubing; ☐ other ______ Sill cocks, number ______
Domestic water heater: type ______; make and model ______; heating capacity ______
______ gph. 100° rise. Storage tank: material ______; capacity ______ gallons.
Gas service: ☐ utility company; ☐ liq. pet. gas; ☐ other ______ Gas piping: ☐ cooking; ☐ house heating.
Footing drains connected to: ☐ storm sewer; ☐ sanitary sewer; ☐ dry well. Sump pump; make and model ______
______; capacity ______; discharges into ______

23. HEATING:

☐ Hot water. ☐ Steam. ☐ Vapor. ☐ One-pipe system. ☐ Two-pipe system.
☐ Radiators. ☐ Convectors. ☐ Baseboard radiation. Make and model ______
Radiant panel: ☐ floor; ☐ wall; ☐ ceiling. Panel coil: material ______
☐ Circulator. ☐ Return pump. Make and model ______; capacity ______ gpm.
Boiler: make and model ______ Output ______ Btuh.; net rating ______ Btuh.
Additional information: ______
Warm air: ☐ Gravity. ☐ Forced. Type of system ______
Duct material: supply ______; return ______ Insulation ______, thickness ______ ☐ Outside air intake.
Furnace: make and model ______ Input ______ Btuh.; output ______ Btuh.
Additional information: ______
☐ Space heater; ☐ floor furnace; ☐ wall heater. Input ______ Btuh.; output ______ Btuh.; number units ______
Make, model ______ Additional information: ______
Controls: make and types ______
Additional information: ______
Fuel: ☐ Coal; ☐ oil; ☐ gas; ☐ liq. pet. gas; ☐ electric; ☐ other ______; storage capacity ______
Additional information: ______
Firing equipment furnished separately: ☐ Gas burner, conversion type. ☐ Stoker: hopper feed ☐; bin feed ☐
Oil burner: ☐ pressure atomizing; ☐ vaporizing ______
Make and model ______ Control ______
Additional information: ______
Electric heating system: type ______ Input ______ watts; @ ______ volts; output ______ Btuh.
Additional information: ______
Ventilating equipment: attic fan, make and model ______; capacity ______ cfm.
kitchen exhaust fan, make and model ______
Other heating, ventilating, or cooling equipment ______

24. ELECTRIC WIRING:

Service: ☐ overhead; ☐ underground. Panel: ☐ fuse box; ☐ circuit-breaker; make ______ AMP's ______ No. circuits ______
Wiring: ☐ conduit; ☐ armored cable; ☐ nonmetallic cable; ☐ knob and tube; ☐ other ______
Special outlets: ☐ range; ☐ water heater; ☐ other ______
☐ Doorbell. ☐ Chimes. Push-button locations ______ Additional information: ______

25. LIGHTING FIXTURES:

Total number of fixtures ______ Total allowance for fixtures, typical installation, $ ______
Nontypical installation ______
Additional information: ______

DESCRIPTION OF MATERIALS

26. INSULATION:

Location	Thickness	Material, Type, and Method of Installation	Vapor Barrier
Roof			
Ceiling			
Wall			
Floor			

HARDWARE: *(make, material, and finish.)*

SPECIAL EQUIPMENT: *(State material or make, model and quantity. Include only equipment and appliances which are acceptable by local law, custom and applicable FHA standards. Do not include items which, by established custom, are supplied by occupant and removed when he vacates premises or chattles prohibited by law from becoming realty.)*

27. MISCELLANEOUS: *(Describe any main dwelling materials, equipment, or construction items not shown elsewhere; or use to provide additional information where the space provided was inadequate. Always reference by item number to correspond to numbering used on this form.)*

PORCHES:

TERRACES:

GARAGES:

WALKS AND DRIVEWAYS:

Driveway: width ______ ; base material ______ ; thickness ______"; surfacing material ______ ; thickness ______"

Front walk: width ______ ; material ______ ; thickness ______". Service walk: width ______ ; material ______ ; thickness ______"

Steps: material ______ ; treads ______"; risers ______". Cheek walls ______

OTHER ONSITE IMPROVEMENTS:

(Specify all exterior onsite improvements not described elsewhere, including items such as unusual grading, drainage structures, retaining walls, fence, railings, and accessory structures.)

LANDSCAPING, PLANTING, AND FINISH GRADING:

Topsoil ______" thick: ☐ front yard; ☐ side yards; ☐ rear yard to ______ feet behind main building.

Lawns *(seeded, sodded, or sprigged)*: ☐ front yard ______ ; ☐ side yards ______ ; ☐ rear yard ______

Planting: ☐ as specified and shown on drawings; ☐ as follows:

______ Shade trees, deciduous, ______" caliper.

______ Low flowering trees, deciduous, ______' to ______'

______ High-growing shrubs, deciduous, ______' to ______'

______ Medium-growing shrubs, deciduous, ______' to ______'

______ Low-growing shrubs, deciduous, ______' to ______'

______ Evergreen trees. ______' to ______', B & B.

______ Evergreen shrubs. ______' to ______', B & B.

______ Vines, 2-year ______

IDENTIFICATION.—This exhibit shall be identified by the signature of the builder, or sponsor, and/or the proposed mortgagor if the latter is known at the time of application.

Date ______ Signature ______

Signature ______

Glossary

The following management, financial, building, and pool[1] terms may be of assistance to you.

Algae: A microscopic plant life that thrives and multiplies very rapidly, especially in warm, unchlorinated water. Algae cause slimy patches and stains to develop on the bottom and sides of the pool. There are many stains of algae, but the most common are the green, reddish brown, or black.

Assets: What an individual or organization owns plus what is owed to it. Assets may be tangible, like a house, or intangible, like good will.

Available Chlorine: A measure of active chlorine present in your pool water to combat germs and algae.

Balanced Water: Pool water that is chemically balanced; that is to say water that has a pH reading of between 7.2 and 7.5. Water that is neither too alkaline nor too acid.

Balance Sheet: A financial report on an individual or organization, listing their assets, liabilities, and net worth for one particular period at a time.

Bond: A formal evidence of a debt, in which the borrower agrees to pay the lender a specified amount, with interest at a fixed rate payable on specified dates.

Budget: A schedule for estimating expenses for a given period. It can be used by management as both a planning and control document.

Capital Structure or Capitalization: The amount and type of securities authorized and issued by an organization.

Cash Items: Cash, bank deposits, U.S. government issues, and other securities that are considered the same as cash.

Caveat Emptor: "Let the buyer beware."

Collateral: An obligation or security attached to another to secure its performance.

Common Stock: Securities which have a right to dividends subordinate to all other stock of the organization.

Communication: The giving and receiving of information.

Conveyance: An instrument by which title to property is conveyed.

Coordination: Effective communication among members of an organization to permit them to function in harmony.

Current Assets: Cash, cash items, inventories, notes and accounts receivable due within one year.

[1] Source for the pool terms is *Pool Life*, Olin Corporation, Stamford, Connecticut, 1973, p. 3.

Current Liabilities: Accounts and notes payable; accrued taxes; interest; declared dividends; and other claims that are payable within one year.

Decision-making Process: The decision-making process may draw upon the following management components in order to arrive at a sound solution: objective, resources, areas, and functions. In arriving at management decisions, these questions should be kept in mind: (1) What is your objective? (2) Do you have the necessary facts to make a sound decision? (3) What are your alternatives? and (4) Have you chosen the most profitable alternative?

Depreciation: A periodic charge against income to spread the cost of such items as building and equipment over their estimated useful life.

Equity: The net worth of an organization that presents the money value owned by the stockholders.

Feedback: The receipt of information by management obtained both from within and outside an organization. A primary means of effective feedback is through conversation with employees.

First Mortgage: A mortgage having precedence over all others.

First Mortgage Bond: A bond secured by a first mortgage on property of the issuing corporation.

Floater: A device used on the surface of your pool water for dispensing dry chlorine in tablet form.

Floor Joists: Floor framing lumber which extends from the outer foundation walls to interior beams.

Footings: The concrete base of a foundation wall.

Foreign Matter: Materials such as dust, twigs, grass clippings, and algae spores, carried into the pool by wind, rain, and bathers. They may carry bacteria and algae which would increase consumption of dry chlorine.

Foundation Wall: A masonry wall supporting the house.

Human Understanding: Respect and compassion for people in an organization.

Incentives: A technique utilized to motivate individuals in an organization.

Leader: The "boss," manager, or head of any organization.

Liabilities: What an individual or organization owes.

Line: An organizational structure in which the superior-subordinate relationship is delineated.

Maintenance: Work required to keep equipment operating.

Management: The art and science of supervising an organization. Also, the supervisory members of an organization.

Manpower: The total output of labor in an activity to achieve an objective.

Millwork: Doors, trim, shelving, window sills, and other finishing work.

Mortgage: A giving of property as security for payment of a debt.

Motivate: Encourage individuals in an organization to improve their productivity.

Net Working Capital: Current assets minus current liabilities.

Net Worth: The true financial worth of a person or organization. It can be determined by subtracting liabilities from assets.

Organizational Chart: A schematic drawing portraying the formal relationships among the various activities of an organization.

PERT (Program Evaluation and Review Technique): It serves as a manager's tool for defining and coordinating what must be done to successfully accomplish the objectives of a project within an established time frame.

Activity: The actual performance of a task. It is the time-consuming portion of the PERT network and requires the resources of men, money, and material.

Critical Path: The sequence of events that indicates the minimum time required to complete a project.

Event: The start or completion of a task.

T_E: Earliest Completion Time. This is the longest time-consuming path.

pH: A numerical rating to indicate acid or alkaline condition of water. pH 7 is neutral. A rating over 7 is alkaline, under 7 is acid.

PPM: An abbreviation of "parts per million." It is applied to pool water ratings as the quantity of residual chlorine per million parts of water.

Second Mortgage: An additional mortgage placed on property already encumbered by a first mortgage.

Sheathing: The first covering of boards or waterproofing material on the outside wall of a frame house.

Skimmer: A metal or plastic screen to remove debris from the pool water. Can be either permanently built into the wall of an in-ground pool or can be a simple device attached to the intake line of the filter of an above-ground pool. Some pool owners use a manual skimmer which is a net-like device attached to the end of a pole.

Span of Control: The number of people in an organization reporting to one supervisor.

Speculation: A home built for the purpose of selling it to anyone willing to pay the price. This is in contrast to a custom built house that has been contracted for prior to construction.

Stabilizer (conditioner): A special chemical agent (cyanuric acid) which, when applied to pool water in recommended amounts, slows the dissipation rate of the chlorine residual. (Especially useful in warmer climates.)

Statement of Income: A financial report of an individual or organization listing their income, expenses, and profit or loss for a given period, normally one year.

Studs: Horizontal lumber nailed to vertical lumber which comprises the wall frame. They are usually 16 inches apart.

Subfloor: Wood sheeting nailed to floor joists.

Superchlorination: Generally referred to as "shock treatment," adding two or three times the normal amount of chlorine to the water in your pool.

Suspended Matter: Particles that do not settle in the bottom and give a cloudy or milky appearance to the water.

Terrazzo: A mosaic flooring made by embedding small pieces of marble or granite in marble and polishing.

Bibliography

The publications listed here will be of interest to those who desire additional background on management and homebuilding. The articles and books are divided into ten parts to correspond with the chapters. Much of this material can be found in the larger libraries.

Chapter 1: A Managerial Approach to Building Your Home

Books:

Anderson, LeRoy Oscar, and Zornig, Harold F. *Build Your Own Low Cost Home.* New York: Dover Publications, 1972.

Dale, Ernest. *Management: Theory and Practice,* Part 1, Management and Its Environment. 3rd ed. New York: McGraw-Hill, 1973.

Drucker, Peter F. *The Practice of Management,* Introduction: The Nature of Management. New York: Harper & Row, 1954.

Eisinger, Larry. *How to Build and Contract Your Own Home,* rev. ed. Greenwich, Conn.: Fawcett Publications, 1971.

Higson, James D. *The Higson Home-Builder's Guide.* Los Angeles: Nash Pub., 1972.

Koontz, Harold, and O'Donnell, Cyril. *Principles of Management.* Part 1, The Basis of Management. New York: McGraw-Hill Book Co., 1972.

Neal, Charles. *Do-It-Yourself Housebuilding: Step-by-Step.* New York: MacMillan, 1973.

Schuler, Stanley. *All Your Home Building and Remodeling Questions Answered.* New York: Macmillan, 1971.

Stillman, Richard J. *Guide to Modern Management.* Part 1, Purpose and Scope of Modern Management. Englewood Cliffs, N.J.: Prentice-Hall, forthcoming.

Articles:

Dempewolff, R. F. Building a House? 20 Tips That Save Time and Your Back. *Popular Mechanics.* May 1971. pp. 112-115ff.

Drucker, Peter F. The Effective Decision. *Harvard Business Review.* January-February 1967. pp. 92-98.

Ingersoll, J. H. We're Building What We Like; S. Jacob and T. Rice of Plainfield, Vt. *House Beautiful.* July 1970. pp. 50-51ff.

Schuler, S. House Building: Dream or Nightmare. *American Home* May 1970. pp. 40ff.

Chapter 2: Planning Your Dream House

Books:

Dale, Ernest. *Readings In Management: Landmarks and New Frontiers.* Chapter 16, Planning and Forecasting. 2d ed. New York: McGraw-Hill, 1970.

Donnelly, James H., Jr.; Gibson, James L.; and Ivancevich, John M. *Fundamentals of Management.* Chapter 4, The Planning Function. Austin, Texas: Business Publications, Inc., 1971.

Federal Housing Administration. *Minimum Property Standards for One and Two Living Units.* No. 300, 1966.
Haynes, W. Warren, and Massie, Joseph L. *Management: Analysis, Concepts, and Cases,* 2d ed. Englewood Cliffs, N.J.: Prentice-Hall, 1969.
Miner, John B. *The Management Process.* Part 2, Planning: The Establishment of Role Prescriptions. New York: Macmillan, 1973.
U.S. Department of Agriculture, *House Construction (How to Reduce Costs),* Bulletin No. 168. Washington, D.C., August 1969.

Articles:
Miller, R. W. How to Plan and Control With PERT. *Harvard Business Review.* March-April 1962, pp. 93-104.
Schneider, Frank L. Return to the City Beginning Here? Some Feel It May Be. *New Orleans Times-Picayune.* February 18, 1973, p. 17.
Tomb, J. O. A New Way to Manage: Integrated Planning and Control. *California Management Review.* Fall, 1962, pp. 57-62.

Chapter 3: Money—The Essential Ingredient

Books:
Hodge, Billy J., and Johnson, Herbert J. *Management and Organizational Behavior: A Multidimensional Approach.* Chapter 17, Measuring Mission Accomplishment Through Financial Analysis. New York: Wiley, 1970.
Scanlan, Burt K. *Principles of Management and Organizational Behavior.* Chapter 24, Accounting and Financial Controls. New York: Wiley, 1973.
Stillman, Richard J. *Guide to Personal Finance: A Lifetime Program of Money Management.* Chapter 4, Housing. Englewood Cliffs, N.J.: Prentice-Hall, 1972.
Veterans Administration. *Pointers for the Veteran Homeowner.* rev. ed. VA Pamphlet 26-5. Washington, D.C., May 1971.
Veterans Administration. *Questions and Answers On Guaranteed and Direct Loans for Veterans.* rev. ed. VA Pamphlet 26-4. Washington, D.C., May 1971.
Veterans Administration. *To the Home-Buying Veteran.* rev. ed. VA Pamphlet 26-6. Washington, D.C., May 1971.

Articles:
Argyris, Chris. Human Problems With Budgets. *Harvard Business Review.* XXXI (1), pp. 97-110.
Hughes, Charles L. Why Budgets Go Wrong. *Personnel* vol. 42, no. 3, May-June 1965, pp. 19-26.

Chapter 4: Building Your Home

Books:
Hicks, Herbert G. *The Management of Organizations: A System of Human Resources Approach.* Section C, Organizing. 2d ed. New York: McGraw-Hill, 1972.
Massie, Joseph L., and Douglas, John. *Managing: A Contemporary Introduction.* Chapter 6, Organization Structure. Englewood Cliffs, N.J.: Prentice-Hall, 1973.
McFarland, Dalton E. *Management: Principles and Practices.* Part III, Organization. New York: Macmillan, 1970.
Sisk, Henry L. *Principles of Management.* Part III, The Organizing Function. Cincinnati, Ohio: South Western Publishing Company, 1969.

Articles:
Insulation More Important Now Than Ever, *Homemaking.* March 1973, New Orleans Public Service, Middle South Utility System.

Koontz, H. Making Theory Operational: The Span of Management. *The Journal of Management Studies*. October 1966, pp. 229-243.
Stieglitz, H. Optimizing Span of Control. *Management Record*. September 1962, pp. 25-29.
Wong, William. Whopping Price Hikes on Lumber Products Wallop Home Builders. *Wall Street Journal*. March 5, 1973, p. 1.

Chapter 5: CHECKING ON WHAT YOU HAVE DONE

Books:

Haynes, W. Warren, and Massie, Joseph L. *Management: Analysis, Concepts, and Cases*. 2d ed. Chapter 13, Control. Englewood Cliffs, N.J.: Prentice-Hall, 1969.
Longenecker, Justin G. *Principles of Management and Organizational Behavior*. 2d ed. Controlling Organizational Performance. Columbia, Ohio: Charles E. Merrill, 1969.
Newman, William H.; Summer, Charles E. and Warren, E. Kirby. *The Process of Management*, 3d ed. Part Six, Measuring and Controlling. Englewood Cliffs, N.J. Prentice-Hall, 1972.
Scanlan, Burt K. *Principles of Management and Organizational Behavior*. Chapter 22, Managerial Control. New York: Wiley, 1973.

Articles:

Sherwin, Douglas S. The Meaning of Control. *Dun's Review and Modern Industry*. January 1956, pp. 45, 46, and 83.
Tannenbaum, Jeffrey A. Homeowners Outraged by New House Defects and Delays in Repairs. *Wall Street Journal*. April 3, 1973, p. 1.

Chapter 6: DO YOU WANT A SWIMMING POOL?

Books:

Butler, George D. *Outdoor Swimming Pools; Considerations in Planning, Basic Design Features, Pool Construction Factors*. New York: National Recreation Association. 1955.
Ideal Home Magazine. *Patios and Pools*. London, New York: Hamlyn, 1969.
Joseph, James. *Poolside Living*. Garden City, N.Y.: Doubleday, 1963.
Pedersen, H. W. *Pool Owners Handbook*. Miami Shores, Florida, 1969.
Springer, John L.; Hochman, Louis; and Gladstone, Bernard. *All About Swimming Pools*. Greenwich, Conn.: Fawcett, 1960.
Sunset Magazine. *Swimming Pools*. 4th ed. Menlo Park, Calif. Lane Books, 1970.
Wills, Norman D. *Build Your Own Swimming Pool: A Guide to Pool Care and Construction*. London: Gifford, 1969.

Articles:

Gas-Lit Torches Around Their Pool. *Sunset*. July 1971, p. 100.
How to Keep Your Pool Crystal Clear. *Popular Mechanics*. June 1971, pp. 154-157.
LaGregga, F. What It Really Costs to Maintain a Swimming Pool. *Mechanix Illustrated*. August 1971, p. 63.
Pool Enclosures: From Domes to Greenhouses. *Business Week*. August 28, 1971, p. 67.
Pool Life Magazine. Olin Corp., Stanford, Conn., seasonal.

Chapter 7: PROTECTING YOUR PROPERTY

Books:

Arnold, Peter. *Burglar-Proof Your Home and Car*. Los Angeles: Nash Pub., 1971.
Arnold, Robert Taylor. *The Burglars Are Coming*. Santa Ana, Calif.: Arnold Pub., 1972.
Brann, Donald R. *How To Install Protective Alarm Devices*. Briarcliff Manor, N.Y.: Directions Simplified, 1972.

Cunningham, John Edward. *Security Electronics.* Indianapolis: H. W. Sams, 1970.
Eagle, Alyn Bush. *Family and Home Protection Against Crime.* College Park, Md.: Executive House, Services, 1968.
Healy, Richard J. *Design for Security.* New York: Wiley, 1968.
Kaufmann, Ulrich George. *How to Avoid Burglary, Housebreaking, and Other Crimes.* New York: Crown, 1967.
Sootin, Hy. *Burglar Alarm Security.* Miami: Sootin's, 1970.
Treves, Ralph. *Do-It-Yourself Home Protection; A Common-Sense Guide.* New York: Popular Science Pub. Co., 1972.

Articles:
Day, R. How to Protect Your Home With Lights. *Mechanix Illustrated.* March 1971, pp. 120ff.
Giving Robbers the Cold Shoulder. *Today's Health.* February 1971, p. 41.
Here's a Plan Thieves Don't Like: Operation Identification. *Better Homes and Gardens.* March 1971, p. 126.
Hill, R., and Hawkins, W. J. Do-It-Yourself Guide to Making Your Home Secure. *Popular Science.* August 1972, pp. 94-97.
Ingersoll, J. H. Security For Your Home; More Urgent Than Ever. *House Beautiful.* November 1971, pp. 62ff.
Lee, A. Ways to Protect Your Home. *Better Homes and Gardens.* June 1972, pp. 16ff.
Protect Your Home Against Burglars. *Good Housekeeping.* September 1972, p. 187.
Rush, A. F. This Man Can Be Stopped! *McCalls.* March 1972, pp. 86-87ff.
Some Minimum Security Precautions. *Consumer Report.* February 1971, p. 103.
Torok, L. How You Help the Burglar. *Popular Mechanics.* September 1971, pp. 86-87.

Chapter 8: How Too Keep Your Cool While Moving

Books:
Giammattei, Helen, and Slaughter, Katherine. *Help Your Family Make a Better Move.* rev. ed. New York: Doubleday, 1970.
Randall, Margaret. *The Home Encyclopedia of Moving Your Family.* New York: Berkley Publishing, 1959.
Rosefsky, Robert S. *The Ins and Outs of Moving.* Chicago: Follett Publishing, 1972.
Sullivan, George. *Do-It-Yourself Moving.* New York: Macmillan, 1973.
Warmington, Carl. *The Family Guide to Successful Moving.* New York: Association Press, 1968.

Articles:
Daly, M. Up-to-Date Tips About Moving. *Better Homes and Gardens.* July 1972, pp. 20-21ff.
DeMarco, J. How to Move Yourself. *Mechanix Illustrated.* February 1971, pp. 74-75ff.
Easing the Trauma of Moving. *Business Week.* March 13, 1971, p. 101.
Hand, J. How to Get an Honest Move. *Mechanix Illustrated.* June 1971, pp. 67-68.
Irwin, T. It's Legal; Question of Lost or Damaged Items. *House Beautiful.* September 1972, pp. 20ff.
Rush, A. F. Taking the Terror Out of Moving. *McCalls.* March 1971, pp. 35-37.
Sure You Can Do Your Moving Yourself. *Changing Times.* September 1971, pp. 35-37.
Weinstein, G. W. Moving Made Easy. *Parents Magazine.* February 1971, pp. 70-71.
When Moving Men Show Their Worst Side. *Business Week.* August 21, 1971, p. 86.

Chapter 9: Landscaping

Books:
Ajay, Betty. *Betty Ajay's Guide to Home Landscaping.* New York: McGraw-Hill, 1970.

Brooks, John. *Room Outside; A Plan for the Garden.* New York: Viking Press, 1970.
Bruning, Walter F. *Home Garden Magazine's Minimum Maintenance Gardening Handbook.* New York: Harper & Row, 1970.
Flemer, William, III. *Nature's Guide to Successful Gardening and Landscaping.* New York: Crowell, 1972.
Kramer, Jack. *Gardening and Home Landscaping Guide.* New York: Arco, 1970.
Macleod, Dawn. *Design Your Own Garden.* London: Duckworth, 1969.
Phillips, Cecil Ernest Lucas. *The Design of Small Gardens.* London: Heinemann, 1969.

Articles:
Mason, H. Relaxing Solutions for Landscaping Problems. *Better Homes and Gardens.* March 1971, pp. 46-55.
Planting for Privacy. *Sunset.* December 1971, pp. 198-199.
Planting Ideas for Side Yards. *Sunset.* July 1971, p. 161.
Rocks and Plants: Good Companions. *Sunset.* April 1972, pp. 261-262.
Smith, A. U. Cool It With Trees. *Horticulture.* August 1971, pp. 18-21.

Chapter 10: MAINTENANCE, REPAIRS, IMPROVEMENTS

Books:
Audel, T. *Audels Do-It-Yourself Encyclopedia.* Indianapolis: How-To Associates, 1968.
Cassiday, Bruce. *The New Practical Home Repair for Women: Your Questions Answered.* New York: Taplinger Pub. Co., 1972.
Cobb, Hubbard, *How to Paint Anything.* New York: Macmillan, 1972.
Cobb, Hubbard, *Money Saving Home Maintenance.* New York: Collier Books, 1970.
Curry, Barbara A. *Okay, I'll Do It Myself; or A Handy-Woman's Primer That Takes the Mystique Out of Home Repairs.* New York: Random House, 1971.
Evans, Melvin. *Easy Home Repairs.* New York: Western Pub. Co., 1972.
Family Handyman Magazine. *The Family Handyman.* rev. ed. New York: Scribner, 1970.
Sara, Dorothy. *Home Fix-It Encyclopedia; Complete Manual of Home Repairs and Maintenance.* New York: Westport Corp., 1971.
Schuler, Stanley, *The Homeowner's Minimum-Maintenance Manual.* New York: M. Evans, 1971.
Sunset Books and Magazines, eds. *Basic Home Repairs Illustrated.* Menlo Park, Calif.: Lane Books, 1971.
Watkins, Arthur Martin. *The Homeowner's Survival Kit: How to Beat the High Cost of Owning and Operating Your Home.* New York: Hawthorn Books, 1971.
Williamson, Dereck. *The Complete Book of Pitfalls; A Victim's Guide to Repairs, Maintenance, and Repairing the Maintenance.* New York: McCall, 1971.

Articles:
Easy Home Repairs. *Mechanix Illustrated.* August 1971, pp. 80-82.
Home Maintenance Ideas. *Better Homes and Gardens.* May 1972, pp. 22ff.
How to Get Your Money's Worth on Home Improvements. *Good Housekeeping.* November 1970, pp. 168-169.
How to Save on Home Maintenance Costs. *Good Housekeeping.* January 1970, p. 140.
Is Your Home Ready for Winter. *McCalls.* November 1972, pp. 24ff.
Master Plan for a Maintenance-Free Summer. *Better Homes and Gardens.* April 1972, pp. 22ff.
McWhirter, W. A. Horrors of Home Repair; Unskilled and Inaccessible Craftsmen. *Life.* June 1970, pp. 58-60ff.
O'Brien, R. Twelve Money-Saving Tips on Maintaining Your Home. *Reader's Digest.* June 1970, pp. 189-190ff.
Three Good Home Repair Tips. *Popular Science.* October 1970, p. 127.
Twenty Great Home Fix-Up Ideas. *Mechanix Illustrated.* February 1970, pp. 83-93.

Index

Nos. in **boldface** refer to illustrations